IMPLEMENTING CONCURRENT PROJECT MANAGEMENT

IMPLEMENTING CONCURRENT PROJECT MANAGEMENT

Quentin C. Turtle
University of Rhode Island
and Technology Management Group

PTR Prentice Hall
Englewood Cliffs, New Jersey 07632

Library of Congress Cataloging-in-Publication Data
Turtle, Quentin C.
    Implementing concurrent project management / Quentin C. Turtle.
      p.   cm. —(Engineering and systems management)
    Includes bibliographical references and index.
    ISBN 0-13-302001-0
    1. Production engineering.   2. Concurrent engineering.
    3. Industrial project management.   I. Title.   II. Series.
    TS176.T88 1994
    658.5—dc20   93-2140
                        CIP

Editorial/production supervision: *Kerry Reardon*
Cover design: *DeFranco Design, Inc.*
Manufacturing manager: *Alexis Heydt*
Acquisitions editor: *Michael Hays*

©1994 by PTR Prentice Hall
Prentice-Hall, Inc.
A Paramount Communications Company
Englewood Cliffs, New Jersey 07632

The publisher offers discounts on this book when ordered
in bulk quantities. For more information, contact:

Corporate Sales Department
PTR Prentice Hall
113 Sylvan Avenue
Englewood Cliffs, NJ 07632

Phone: 201-592-2863
Fax: 201-592-2249

Printed in the United States of America

10 9 8 7 6 5 4 3 2

ISBN   0-13-302001-0

Prentice-Hall International (UK) Limited, *London*
Prentice-Hall of Australia Pty. Limited, *Sydney*
Prentice-Hall Canada Inc., *Toronto*
Prentice-Hall Hispanoamericana, S.A., *Mexico*
Prentice-Hall of India Private Limited, *New Delhi*
Prentice-Hall of Japan, Inc., *Tokyo*
Simon & Schuster Asia Pte. Ltd., *Singapore*
Editora Prentice-Hall do Brasil, Ltda., *Rio de Janeiro*

Contents

PREFACE

My goal in writing *Implementing Concurrent Project Management* has been to provide a complete procedure for implementing concurrent engineering in the planning, scheduling, and controlling of technical projects.

The primary objective is to provide the reader with a detailed procedure for concurrent project management by first laying a background definition of and motivation for the relatively new discipline of concurrent project management and then presenting a procedure for the planning, scheduling, and controlling of projects implementing this new discipline.

For several years now, I have taught a course in technical project management at the University of Rhode Island. This course is based on the discipline of concurrent engineering. What is new about this course? Technical project management courses have been around for many years. There have been numerous seminars on concurrent engineering, sometimes called simultaneous engineering or synchronous engineering. However, there are few, if any, three credit courses taught at the university level that cover the traditional, strongly required elements of technical project management, such as planning, scheduling, and controlling, *and* which also are structured around and based on *concurrent engineering*.

Let us analyze, now, the components of the foregoing paragraph.

Concurrent engineering is the carrying out, during the course of the total development of a new product within a company, of all the various engineering disciplines concurrently. The research engineer, the marketing manager, the product engineer, the manufacturing engineer, and the quality engineer all perform as a team in developing the new product.

It is of utmost importance that

1. the manufacturing processes required for producing the new product be developed and tested *concurrently* with the design and development of the new product itself;

2. the proper test procedures and test equipment be qualified *concurrently* with the design and development of the new product itself;

3. the marketing and financial requirements be met *concurrently* with the design of the product and with the design of the manufacturing requirements.

Emphasis must be placed on the whole cross-functional team buying-in. The marketing product manager is an important member of the team. It is important to maintain a strong focus on the needs and wishes of the customer from product concept to customer satisfaction with the product after purchase. The whole point is to gain company-wide concurrence.

Therefore, all of the essential functions for new product development and distribution are carried out *concurrently.*

The term *technical project management* is another component of the foregoing paragraph. This term is to be interpreted as a very comprehensive, critically defined, important segment of management within a company developing new products. For purposes of illustration, can you imagine if the management of the new product development is not treated very critically? Let us assume that a very traditional but very nonchalant attitude were taken toward new product development. Let us work hard at the new product development, but fail to properly address modern-day competitive requirements. What you will get is what you got when it was done that way ten years ago. No, we must do better. We must be modern and competitive, for the companies who are modern and competitive succeed over those who are not. Conferences, seminars, workshops, and other meetings are held to stress these matters. Well-run companies pay great attention to new product development management.

Technical project management means the superb management of the comprehensive project of developing a highly technical new product. It means the proper addressing of all the issues concurrently. It may be a small product, or it may be a large system involving complex software and numerous system components. It may be a small, purely mechanical product, or it may be a large system of components embodying electrical, mechanical, electronic, and hydraulic aspects—all under microprocessor, distributed-processor control. In any case, most new products under development today are high technology products. They may be small products but to be competitive they must use advanced materials, for example. Or, it may be the case that they must be CAD-CAM designed. That is, they embody forward-thinking concepts and require modern, competitive approaches, materials, and software computer technology. Therefore, they are high technology products. This book is concerned with the management of projects wherein such products are developed.

The word *project* is to be interpreted as the carrying forth of the overall development of the product. Overall development of a product includes all the activities, beginning with the concept and proceeding through the production of the product until it is flowing smoothly through the production facility at a good profit.

The term *technical project management* is thus defined by the foregoing paragraphs.

Also, it should be stated at this point, as it will at intervals throughout this book, that the principles to be set forth apply at any level of management. Certainly, the project manager, an individual whose function will be more clearly defined in this book, will benefit from practicing these principles. In addition, the individual team member will greatly benefit. Every individual, at whatever level, must manage his or her time most effectively. Therefore, these principles apply at all levels.

In summary, this book has the following features:

1. It is based on concurrent engineering, which implies that a cross-functional team designs and develops manufacturing processes and facilities and quality assurance processes and facilities concurrently with the design of the product itself.

This book provides the reader with the basis for total quality management (TQM) in product development. TQM in project management means quality in management at all levels by everyone involved with the project. It will become apparent to the reader that the whole content of the book is an implementation of TQM.

2. A complete procedure for implementing concurrent project management is detailed and presented.

3. The project schedule is computed for a specified probability, as well as for the critical path schedule, which is only 50 percent probable!

4. The book is based on a three-credit college course and, yet, it is useful for anyone in college or in industry. When this material has been used, there has been a mix of students from industry and college. The book is designed both as a text for people taking a course and as a training and reference manual for people on the job who wish to improve their skills who may or may not be taking a formal college course. In either case, *the book provides a maximum of understanding in a minimum of time.*

5. The book provides a text suitable for a workshop/seminar type of atmosphere in which the readers participate in three projects.

6. The book reflects an expertise in the subjects covered that has been achieved by the author, who has spent in excess of twenty-five years in industry in new product development.

This book is a product of many experiences of a wide variety as a product development manager in three major industries. It is also a product of experiences gained from speaking at many conferences, teaching, and dialoguing with students from industry over the last four years, and consulting in industry over a long period of time.

There is a continuing urgency in new product development organizations to improve the productivity of the people involved, and thereby to reduce the time to market, the program cost, and the product cost through better design, and to improve the product quality. In the course of treating these matters, competition has become even stronger. Through these exigencies there has been an even stronger motivation for training people toward improved skills. This book provides a rapid, highly efficient means for training people to achieve these results.

The reader must have the following prerequisite: a keen interest in and motivation for improving skills to be used while a member of a product development team.

Chapter 1 introduces and defines the concurrent project management discipline. The psychology of human behavior, cross-functional teamwork, and consensus management are discussed. The reader is referred to the Table of Contents for detail. The overall procedure used throughout the book is also presented.

Chapter 2 covers concurrent project planning.

Chapter 3 covers concurrent project scheduling, Part I, which is a detailed presentation of the network diagrams representing typical projects.

Chapter 4 covers concurrent project scheduling, Part II, which is a discussion of resource allocation including work breakdown structure.

Chapter 5 is a detailed discussion of development cost and how to generate a cost schedule for a project.

Chapter 6 covers reporting, including preparation of the proposal.

Chapter 7 presents the technique for planning, scheduling, and controlling multiple projects.

Chapter 8 is a detailed discussion of concurrent project control.

Chapter 9 is a presentation of four case studies.

I am grateful to all those who have made suggestions and otherwise assisted me in this project, including Hermann Viets, dean of the College of Engineering at the University of Rhode Island, now president of the Milwaukee School of Engineering and David Beretta, past chairman of Uniroyal, Inc. and former executive in residence at the College of Business Administration, now president of Amtrol, Inc., both of whom suggested that I write this book; Albert Ondis, chairman and chief executive officer of Astro-Med, Inc.; professors Leland Jackson, Jien-Chung Lo, William Ohley, and Richard Vaccaro at the University of Rhode Island; and my students and colleagues with whom I have worked at the University of Rhode Island College of Continuing Education and in industry for many years. Careful proofreading was contributed by my sister, Joan, and my daughter, Margaret. I am thankful to my daughter Virginia and my sons William and James for their encouragement and support. I also thank my son-in-law Louis and daughter-in-law Jean for their support. I am also deeply indebted to my wife, Bea, who assisted with the typing and proofreading, without whose sacrifice and devotion this book would not have been possible.

Quentin C. Turtle

BIOGRAPHY

QUENTIN C. TURTLE is president of Technology Management Group, a consulting organization in Providence, Rhode Island. He received B.S.E.E. and M.S.E.E. degrees from the University of Connecticut in 1959 and 1961 and went into industry for a number of years. He then returned to academia and studied applied mathematics at Brown University and control systems and computer science at the University of Rhode Island, where he earned a Ph.D. in 1971. He is a registered professional engineer.

Dr. Turtle served with the Esterline Corporation as Vice President of Research and Development of their Federal Products subsidiary for ten years and was in charge of all technology development and transfer. Federal Products markets quality control gauges and instrumentation.

Previously, he was Vice-President of Engineering and Manufacturing at International Data Sciences, a principal supplier of instrumentation in the telecommunications field. Prior to that, he was employed by General Signal Corporation, developing instrumentation for process control.

In his present capacity, Dr. Turtle teaches seminars and courses and consults on technology transfer, engineers' productivity, and project management. A major effort in which he is involved is transferring technology from university labs to the private sector for commercialization. He is also an adjunct professor in the College of Engineering at the University of Rhode Island.

He is a participant in conferences on technology management, and has recently spoken at (1) the Fourth International Conference on "Product Design for Manufacture & Assembly" in Newport, sponsored by the Institute for Competitive Design and Boothroyd-Dewhurst, Inc., 1989; (2) "Simultaneous Engineering: Making it Work" in Detroit, sponsored by the Society of Manufacturing Engineers, 1989; (3) "Managing Multi-Functional Development Teams" in New Orleans, sponsored by the Manufacturing Institute, 1989; (4) "Five Case Studies: The Results of Applying Simultaneous Engineering" in Providence, sponsored by the Technology Council of Rhode Island, 1989; (5) "Using Concurrent Engineering to Develop Specifications That SELL" in Boston, sponsored by IEEE, 1990; and (6) "A New Management Tool for Capturing and Commercializing New Product Technology," The Fifth International Forum on Design for Manufacturability & Assembly in Newport, Boothroyd-Dewhurst, Inc., 1990.

Companies he has worked for and consulted at include BIF Industries, Inc., New York Air Brake Company, General Signal Corporation, International Data Sciences, Inc., Jet Spray, Inc., Harris Intertype Company, Esterline Corporation, Surgilase, Inc., Barlow & Barlow, Ltd., International Technologies, Inc., and Ciba-Corning Diagnostics Corporation.

1 CONCURRENT NEW PRODUCT DEVELOPMENT

INTRODUCTION

A quiet revolution has taken place in the management of technology over the past several years. A whole new product development strategy and implementation has evolved, and a renewed emphasis has been placed on teamwork in product development. This has been largely due to the need to exploit the output from research faster and to take new, higher quality products to the market faster. This is called survival. This quiet revolution is fueled by the need to stay ahead of competition in chosen market niches. This quiet revolution is fanned by the great success realized by those companies that are implementing concurrent engineering and overall concurrent project management in product development.

Many companies have paid heed to the call placed upon them by stronger competition, and they have found ways to improve productivity as well as to shorten the time to the marketplace with products of higher quality and lower cost. Cost, schedule, and quality must all be improved. It is possible to improve all three concurrently.

Most people who study concurrent engineering immediately comprehend the common sense prudence of following such practice.

Concurrent project management enables companies to pay heed to the call, placed upon them by competition, to find ways to improve productivity and to shorten the time to market with products of higher quality and lower cost.

However, tradition is deeply ingrained so most people are set in their ways and do not know how to readily convert to concurrent engineering. People attend seminars and conferences, take courses and discuss concurrent engineering at other professional meetings. They learn that it is also called simultaneous engineering and synchronous engineering, and they learn that those who practice design-for-manufacturability and assembly tend to use the terms synonymously—sometimes in error. They return to their respective companies, discuss the meetings and promote the concept of concurrent engineering. They find too often that the resistance to change is strong, and that inertia in the product development underway—in the older, sequential, traditional way— is too great, and that the forces that cause the older methods to be followed are still there. Yet, there is a continuing success story, told by companies of moment as they convert to and properly implement concurrent engineering. There is *great* success in some cases. This success fans the fire fueled by the critical need to survive in today's global economy. We must drive higher quality products to market faster at lower cost. Simply put, in a company depending on high technology products in a competitive marketplace, we must all address the issue of survival and properly implement modern, product development methods in order to compete well.

People often state: "I know that team building in concurrent engineering is good. We want to follow these practices, but we do not know how. Tell us how to implement these concepts."

The question people have is how to properly implement these new methods. We often hear people at conferences and other professional meetings state: I know that team building in concurrent engineering is good. We want to follow these practices, but we do not know how. Tell us how to implement these concepts.

The question facing everyone setting out to practice concurrent engineering (CE) is which model to implement. They have attended meetings of one sort or another, they have heard the "experts" speak, they sit on subcommittees, and they have discussed the matter at great length. Yet, they do not know where to begin. Even though they have familiarized themselves with case studies in terms of successes in other companies that have applied concurrent engineering, they still do not know how to apply these concepts to *their* new product development in *their* company. The question is still asked: We know we should begin practicing CE, but how do we get started and how do we follow through with it, successfully?

The basic reason for writing this book is to provide an insight into the concepts and discipline of concurrent engineering and, more broadly, concurrent project management. This book will illustrate how to get started in this new discipline and how to successfully follow through with implementing concurrent project management techniques.

This book will illustrate how to get started in this new discipline and how to successfully follow through with it.

We will detail a model in this book, present a detailed procedure, and look at case studies, as well as present several examples of projects, one of which we will carry through the whole process of concurrent project management from beginning to end. We will take this journey together, and at the end you will be experienced and successful. More specifically, you will be better prepared for your future projects in new product development and product line maintenance.

HIERARCHY OF ORGANIZATIONS

It is important for us to orient ourselves properly for addressing the overall subject of concurrent project management and for understanding the various relationships necessary in properly implementing the associated disciplines. Figure 1.1 provides focus in this regard.

We are interested in product development by a cross-functional team. As can be seen from the figure, there are other, higher-level considerations to be addressed, but we will limit our study to product development.

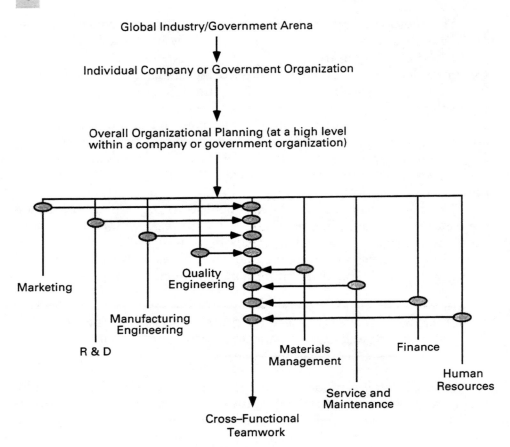

FIGURE 1.1 Hierarchy of organizations and cross-functional teamwork.

At the very top of the figure is the global arena in terms of both industry and international government. We, in fact, live in an era where most products compete globally.

Lower on this scale is the individual private sector company or government organization. Most products are developed by companies. However, in some cases, products are conceived of and developed to a significant extent by government organizations such as the National Institute of Standards and Technology, the Naval Underwater Warfare Center or the Army Research Center. Private sector companies then produce the products under contract. The methods taught in this book apply in either case.

At the next level, as shown in the figure, overall organizational planning takes place at a fairly high level within the company or government organization. The board of directors, for example, selects the top management officers, and they in turn decide that there will be certain product lines. They then define them, arrange for capitalization, and maintain employment levels.

The new product development is carried out by a cross-functional team.

At the functional level, the principal functions, such as marketing, engineering, and manufacturing, operate with financial, human resources, and other support functions in place. The service department, usually under either marketing or engineering, can provide valuable input into the product development process. It is at this operational level that the product cross-functional development team performs.

The cross-functional team is best provided by combining personnel in a functional and matrix organization. That is, the project leader may have a long-term assignment in engineering, for example, and is assigned management of the cross-functional team for the duration of the product development. Other members of the team are drawn from marketing, manufacturing, materials management, and finance. This is illustrated in Figure 1.1 by the linkages with these other functions.

HIERARCHY OF MANAGEMENT AND ENGINEERING LEVELS

There is also a hierarchy in the definitions and interpretations of various disciplines. Figure 1.2 is intended to properly orient the reader when studying the subjects in this book. For example, there has been some confusion in the uses and meanings of the terms *concurrent engineering* and *design-for-manufacturability*. Figure 1.2 will help the reader maintain adequate interpretation in the hierarchy of all the terms. For instance, it is to be understood that concurrent project management probably includes input from marketing, finance, purchasing, and human resources, whereas design-for-manufacturability and engineering component design might not include this input.

Company Management

All the characteristics defined below plus management of company objectives/product mission and profit.

Concurrent Project Management

Everything below plus Marketing, Finance, Purchasing, Human Resources, Engineering, Manufacturing, all in the same team-building process and all achieving concurrence as a team.

Concurrent Engineering

All of the below with characteristics such as team building and gaining concurrence from all team members.

Simultaneous Engineering

DFM and Quality Engineering specification of test procedures/testability design features.

Design-for-Manufacturability(DFM)

Design of the product as driven by manufacturing requirements.

Engineering Component Design

FIGURE 1.2 Definitions and interpretations of various terms and disciplines by heirarchy.

Certainly design-for-manufacturability is a major subset of concurrent engineering which, in turn, is a major subset of concurrent project management. Concurrent project management is defined in this book to include marketing, finance, materials management, human resources and perhaps other functions, as well as engineering. Most often concurrent engineering is used synonymously with simultaneous engineering and is construed by many to mean simply the development of the manufacturing processes simultaneously with the design of the product. In this book, concurrent engineering is given the additional meaning of concurrence by all—consensus engineering. It is in this realm that our competitors on the Pacific Rim excel.

Also, we must remember that the broad meaning of overall concurrent project management, which is a superset of concurrent engineering, should always be kept in sharp focus. Marketing, finance, human resources, and materials management are inputs to the business plan and are of equal rank with research and development, quality assurance, and manufacturing.

DEFINITION OF CONCURRENT PRODUCT DEVELOPMENT

The concept of team-building and consensus management involving simultaneous design, product engineering, manufacturing, and quality engineering and concurrence by all is known as *concurrent engineering*. The Institute for Defense Analysis (IDA) Report R-338 defines concurrent engineering as "A systematic approach to the integrated, concurrent design of products and their related processes, including manufacturing and support. This approach is intended to cause developers, from the outset, to consider all elements of the product life cycle from conception through disposal, including quality, cost, schedule, and user requirements." The concept and implementation of concurrent engineering with the simultaneous involvement of marketing, engineering, manufacturing, finance, and human resources, and the concurrence by all in new product development resulting in successful product life is defined as *concurrent project management*.

The term *concurrent engineering* has been used to describe cross-functional teamwork in new product development. Concurrent engineering is sometimes used synonymously with the term *simultaneous engineering*, which means the simultaneous performance of all the engineering functions in a manufacturing company throughout the new product life cycle. Concurrent engineering/simultaneous engineering provides the product design, the quality engineering, and the manufacturing process engineering all at the same time.

It is crucial for top management and the heads of the departments defined previously to be in complete agreement and that Manufacturing and Quality Assurance be proactive in the conceptual design. Research has revealed that a large percentage of the cost of the new product incurred throughout its life cycle is determined by the characteristics of the product concept. How the functional specifications are met by the conceptual design, as chosen by the cross-functional team, determines the materials of construction and the form and shape of the product. Therefore, the manufacturing processes and methods are set by the conceptual design. Even the manufacturing floor space requirement is decided by the product concept. Floor space is a continuing long-term expense. It is for these reasons that it is crucial for Manufacturing and Quality to have a strong voice in the conceptual design.

EARLY INVOLVEMENT OF MANUFACTURING AND QUALITY ASSURANCE IN DESIGN AND DEVELOPMENT

One can cite many instances where concurrent engineering should be implemented. Whether we are considering new product development or the providing of a "product" in the more general sense of carrying out any project or task, the results of which are to be delivered to other people, examples illustrating the advantages of concurrent engineering abound. It is always better to have cross-functional teamwork, with the manufacturing personnel working with the other principals.

For example, Manufacturing and Quality Assurance (QA) should be involved in the early development to decide whether an instrument housing should be molded plastic, metal casting, or sheet metal weldment. This selection clearly impacts manufacturing processes and this decision should be a manufacturing decision.

Another example is in optical fiber manufacture. As pointed out by I. Magaziner and M. Patinkin in *The Silent War*, Corning Glass has maintained its leadership position in the marketing of optical fibers primarily through manufacturing innovation and implementation of manufacturing leadership concepts.

A third example is the custom chip that worked perfectly in the design lab but could not be manufactured. This is another true case. The chip could not be tested in production; inadequate attention was paid during the design cycle to the production test requirements, and test points were not designed in.

CONCURRENT DEVELOPMENT OF PRODUCTION AND QUALITY PROCESSES

The essence of concurrent engineering is the planning, scheduling, and carrying out of the design of the manufacturing and quality processes concurrently with the design of the product by teamwork and by gaining the concurrence of everyone. It is important for the design and development of the quality engineering and the manufacturing processes to be done concurrently with the design and development of the product. It is also important for the capital equipment for both quality control and manufacturing to be procured earlier than has usually been done in the past. This statement emphasizing concurrent engineering will be made more than once in this book.

Concurrent engineering can be construed to have an additional significant meaning: the concurrence of all people on the team to "buy-in" to good new product development management, excellent design concepts, and productive implementation. When the cross-functional team participates actively and productively from new product concept to customer satisfaction and company profit, the team members all gain concurrence from each other. This is an additional meaning that is significant. This is why we prefer to use the term *concurrent engineering* instead of *simultaneous engineering*.

VARIOUS BUSINESS COMBINATIONS WHEREIN CONCURRENT PROJECT MANAGEMENT IS IMPORTANT

There are various business enterprises in which concurrent engineering and, at a broader level, concurrent project management can be implemented.

Case 1

The most commonly instituted implementation is within one company. The leadership, teamwork, communication, consensus, simultaneous engineering, technology transfer, and other elements composing concurrent project management can take place within one business entity. Marketing, Engineering, Operations, Quality, Finance, and Human Resources all reporting to one president or general manager apply the discipline of concurrent project management. Let us call this Case 1.

Case 2

Alternatively, concurrent project management can be implemented as follows: The supplier of the new product, to its customer base, designs and then develops the design for production by *another* company. For example, instead of the manufacturing operations and the design organization belonging and reporting to one head, there are two separate companies participating in the project management concurrently. This condition makes it no less important or advantageous. In this case, the two separate companies are both represented on the cross-functional team—as though they were all in the same company as in Case 1 previously given. It is extremely important and beneficial for the design of the product in one company and the design and development of the manufacturing processes (as well as the quality engineering) in the other company to be carried out concurrently.

> *Concurrent project management can be implemented in a variety of business combinations.*

Case 3

Case 3 is the situation where one company carries out full concurrent project management on a product that becomes a component part or a subassembly in a larger product. This is similar to Case 1 except that here it is known a priori that the design and production is just a small part of a much larger picture. The producer of the end product and the producer of the smaller component product can and should both benefit from overall leadership, teamwork, communications, consensus, simultaneous engineering, and technology transfer that constitute concurrent project management.

In the case where the project must go out for bid, such as in military projects or construction projects, the winning contractor will be asked to join the cross-functional team after contract signing.

One additional case will be presented before we leave the discussion of various concurrent project management scenarios. This is the case of transferring technology developed in a university laboratory by a research professor to a private sector company for commercialization. This is of increasing importance because there is a vast amount of intellectual property on the campus, and there is increasing interest in the private sector in commercializing this technology. Again, concurrent project management, implemented by both organizations working together, is beneficial.

THE IMPORTANCE OF QUESTIONS ASKED PROPERLY

Consider now all the situations characterizing concurrent project management: early involvement of manufacturing and quality assurance, concurrent development of the product, production processes and quality assurance processes, and various business entities—all working together. In all these situations it is important to ask questions of one another. Each and every individual involved in the product development project needs information. A large database is required to establish and concur on product specifications, resource allocation, cost of product, cost of development, and time schedule. A timely flow of information within the company is needed to guarantee success in materials management, marketing distribution, and finance. Therefore, people need to communicate efficiently, including asking questions.

Questions must be asked in a manner that leads to answers which are meaningful and accurate, relative to the need for asking in the first place. Sometimes answers are true but misleading because of the way the questions were asked. Consider the following situation.

Two priests who, being unsure if it were permissible to smoke and pray at the same time, wrote to the pope for a definitive answer. One priest phrased the question "Is it permissible to smoke while praying?" and was told it was not, since prayer should be the whole focus of one's attention. The other priest asked if it were permissible to pray while smoking and was told that it was, since it is always appropriate to pray.

This example points out that it is possible to obtain two diametrically opposite answers to the same question, depending on how the question is asked.

COMPARISON WITH THE TRADITIONAL PROCESS

Reference is now made to the two charts in Figures 1.3 and 1.4. Figure 1.3 is presented only for comparison and illustrates the older, traditional way in which products were developed without implementing concurrent project management. In our discussion to follow, let us call this the process followed by Company 1.

Figure 1.4 presents the preferred methodology of concurrent project management. The inherent improved quality and the reduced time to market are apparent. The lower product cost and reduced development project cost are also direct results of cross-functional teamwork in the proper implementation of concurrent project management. New thinking is required. *People* develop new products, and before they can practice concurrent engineering they must change. In a later section we will address the psychology of human behavior. A change in mindset is required in order for people to follow the preferred methodology. In the following discussion we call the company following this preferred methodology Company 2.

It is interesting to study two companies: one following the old product development process and one following concurrent project management procedures.

COMPANY-WIDE DEVELOPMENT OF A PRODUCT

THE PEOPLE
Heads of Marketing,
Engineering,
Manufacturing, and
Quality Assurance

Product Manager and
Project Leader

Research and Development Personnel

Identify Product
Study Company's Goals Regarding the Product Line
Define Market Objectives
Study Competition
Market Research—Study Customers' Specific Needs
Write Product Description
Write Functional Specifications
Conceptual Design
Select Project Leader/Product Champion
Study Technical Feasibility
Write Marketing Plan
Study Resources
Construct R & D Prototypes
Test Technical Feasibility
Prepare Prototypes for Market Research
Confirm Suitability for Customers
Determine Acceptable Selling Price
Forecast Sales Volume
Write Preliminary Business Plan
Write Operator's Manual

Week 13

FIGURE 1.3A First phases of development by Research & Development.

(Write Operator's Manual)—(last line of Figure 1.3A)

R & D Personnel

Write Detailed Product Definition
Write Project Description
Week 15   Develop Plan Network
Superimpose Work Breakdown Structure
Superimpose Development Cost Curve
Update Business Plan
Allocate Resources
Design the Product
Design Review Meeting
Procure Components
Week 20   Construct R & D Prototypes
Test R & D Design
Beta Testing
R & D Product Release
Week 30

FIGURE 1.3B Product design by Research & Development.

(R & D Product Release) —(last line of Figure 1.3B)

Manufacturing Engineering (Personnel)

Manufacturing Engineering:
Design & Develop the Manufacturing Processes
Procure Production Tooling
Construct Manufacturing Engineering Prototypes
Make Engineering Changes and Test
Pilot Run

    Quality Assurance (Personnel)
    Identify Quality Plan
    Develop the Quality Engineering Procedures and Equipment
Week 58

      Quality Engineering Tests Design
      Make Engineering Changes and Test
        Production and Distribution

    Week 70

FIGURE 1.3C Continuation of sequential product development.

COMPANY-WIDE DEVELOPMENT OF A PRODUCT

THE PEOPLE

Heads of Marketing,
Engineering,
Manufacturing, and
Quality Assurance

Product Manager and
Project Leader

Cross-Functional
Development Team

Identify Product

Study Company's Goals Regarding the Product Line

Define Market Objectives

Study Competition

Market Research—Study Customers' Specific Needs

Write Product Description

Write Functional Specifications

Conceptual Design (Advanced Manufacturing Engineering active here, also)

Select Project Leader/Product Champion

Study Technical Feasibility

Write Marketing Plan

Study Resources

Construct R & D Prototypes

Test Technical Feasibility

Prepare Prototypes for Market Research

Confirm Suitability for Customers

Determine Acceptable Selling Price

Forecast Sales Volume

Write Preliminary Business Plan

Write Operator's Manual

Week 13

FIGURE 1.4A First phases of development by company-wide team.

15

(Write Operator's Manual)—(last line of Figure 1.4)

Establish Company-Wide, Cross-Functional Development Team
Identify Manufacturing Processes and Manufacturing Requirements
  Identify Quality Plan
Week 15 Write Detailed Product Definition—95 percent final accuracy
    Write Company-Wide Project Description
    Develop Plan Network and Superimpose Work Breakdown Structure and Cost Curve
    Update Business Plan
      Allocate Resources
        Design the Product
        Design and Develop the Manufacturing Processes
        Design and Develop the Manufacturing Facilities
        Design the Quality Engineering Procedures and Equipment
          Procure Components and Production Tool

Week 20

        Construct Company-Wide Design and Development Prototypes
        Cross-Functional Team Tests Design and Processes
        Manufacture Pilot Run of Shipable Units
          Cross-Functional Team Tests Design and Processes
        Beta Testing
        Formal Product Release
          Product Design
          Manufacturing Processes and Facilities
          QA Design and Test Procedures
          Production and Distribution

Week 50

FIGURE 1.4B Concurrent product development.

In order to make a fair comparison of two companies with and without concurrent project management, we assume in the case of both companies that there is an efficient marketing study and business plan at the front end. That is, in this illustration, company goals and marketing objectives are properly addressed in both types of companies. There is sufficient market research, including a study of competitive products. Also, all the inputs to a preliminary business plan are formulated and properly used, at the outset, by both companies. We assume that both companies follow this broad, favorable type of process at the outset of each project. The reason for this assumption is to paint a picture for the reader that sets the stage for a meaningful comparison between the two companies when it comes to considering the new product development process *downstream* from the initial choice of the product and some preliminary market studies. It would not be a meaningful comparison of the conditions downstream if different, unsound policies were practiced in the early stages leading to diverse, impossible projects afterwards.

Some comments must now be made. For example, we know that some new product developments have proceeded without the benefit of proper company goal setting or proper market research. Some companies simply decide somewhere within the organization at some management level that a new product should be developed—it thereby evolves. From an economic standpoint, sometimes it should evolve and sometimes it should not. Occasionally, it happens that an engineer or scientist invents a new product and the conceptual idea is pursued. There is nothing wrong with this process so long as company goals and market conditions are properly studied *before* significant resources are spent on the project. In fact, some very good products have been originated by the engineer or scientist. However, at times a new product idea is pursued too long under less credible circumstances than this. The point is this: Whether the company is marketing-driven or technology-driven, company goals and objectives transcend all other new product development factors and must be addressed at the front end. At any rate, for our purposes here, we assume that the comparison of the two companies bears on the issue of concurrent management and that, regardless of whether concurrent management is in place, both companies otherwise follow good front end business management procedures. It will then be a more meaningful comparison.

Company goals and objectives transcend all other factors and must be addressed at the outset.

So, in both Figures 1.3 and 1.4, the first five detailed activities are the same. However, in Figure 1.3, the illustration of the next processes followed by Company 1 is quite different from that of Company 2, as shown in Figure 1.4. In Figure 1.3A the product description, the conceptual design, and the prototypes are all prepared by an R & D development team.

In Figure 1.4A, the product description, the conceptual design, and the prototypes are developed by a *company-wide* development team.

So far we have just compared the process as depicted by Part A of the figures. An even more extensive difference takes place in Part B of the product development.

In Part B, Company 1 continues the new product development procedure, with the R & D people writing the product description, designing the product, constructing more R & D prototypes, and testing them. Then, when the design passes the tests *in the judqment of the R & D team*, the product is released from R & D to the operations organization. Manufacturing engineering, although in attendance at the design review meeting(s), now assumes a strong responsibility for the product for the first time!

In the old method, the design and development of the manufacturing processes and facilities take place long after the design and development of the product.

The design and development of the manufacturing processes and facilities take place now, long after the design and development of the product. Another set of prototypes is constructed to test the manufacturing processes and the product as it was designed by the R & D people. R & D designed the product without full knowledge of the final manufacturing processes selected for the specific new product. It is here that one or more, and sometimes many, iterations back to R & D for redesign take place. In Company 1, the quality assurance (QA) program is not considered, either, until too late.

As can be seen in Figure 1.3, the QA plan and the development of the quality engineering test design, as well as often-required engineering changes, take place sequentially.

In our illustration, the overall time required to follow Company 1's new product development procedure is seventy weeks. The reader is referred, again, to Figure 1.3.

Let us turn now to Company 2, whose process is illustrated in Figure 1.4. Albeit, still too small a percentage, there is a significant number of companies, small and large, that have found success by following this procedure. Higher quality, lower product cost, and, often more importantly, much faster time to market result from this superior procedure. In the example illustrated, the total product development time is less than one year. Attendant with this more rapid development is a proportionately lower development cost.

We turn now to Figure 1.4B for comparison with the traditional, sequential, more time-consuming and costly, yet less competitive (if really competitive at all) procedure followed by Company 1.

The cross-functional product development team identifies and plans the manufacturing processes, the tooling and other manufacturing requirements, and all the quality engineering and quality assurance requirements early on. The major difference from Company 1 is that the cross-functional team (in which R & D is a strong component) carries out many of the R & D activities, as a team, which in Company 1 are carried out by R & D alone.

In the preferred process, the cross-functional team identifies and plans the manufacturing processes, the tooling and other manufacturing requirements, and all the quality engineering and quality assurance requirements early on.

More fundamentally, (and this also points out the way in which concurrent project management is more comprehensive than concurrent engineering), the product definition, the project description, and the plan network are formulated and detailed by the *team*. The work breakdown structure/resource allocation and cost schedule are provided and validated by the *team*. These five activities take place before any design work is begun. The overall development schedule is more realistic because the resources of the various talents required at various times during the project are more directly represented by various members of the cross-functional team, and these talents are the ones who do the respective estimating. For example, it is known in either the case of Company 1 or Company 2 that a major manufacturing engineering effort will be required before the product is fully developed. Also, in both cases, an overall schedule is required early on for business planning. In Company 1 this is provided by R & D, whereas in Company 2 this schedule is provided by the total cross-functional team. In our example, manufacturing engineering is represented in the team and can more accurately estimate the time required for the manufacturing effort. Estimates of more realistic schedules are far better and more valuable in Company 2. A major reason for schedule slippage, the frequent bane of project management, is that the schedule was unrealistic to begin with. This is avoided in the cross-functional team approach.

A major reason for schedule slippage, the frequent bane of project management, is that the schedule was unrealistic to begin with.

To continue now with the comparison, the product design, the design and development of the manufacturing processes and facilities, and the development of the quality engineering procedures and equipment are all carried out concurrently. They are carried out at the same time *and with the concurrence of everyone on the team.*

- All the members of the cross-functional team do it all together.
- All members of the team *all* buy-in.

In the preferred process, the product design, the design and development of the manufacturing processes and facilities, and the development of the quality engineering procedures and equipment are all carried out concurrently.

The final prototypes are constructed and tested by the cross-functional team. There is no need for several iterations back to R & D for redesign, for everyone is present right there on the team.

In the illustration comparing the two companies, the total time to develop the new product when concurrent project management is properly implemented is less than one year. To be sure, the overall schedule depends on the specific product under development. Electronic instrumentation and hand tools require less time than automobiles and bridges. However, the ratio of Company 2 time spent to Company 1 time spent of 50 to 70 is quite realistic. In many cases the reduction in time in the cross-functional method is even greater.

So we now have a comparison of the recommended procedure with the older, traditional, less-successful procedure. To the extent that those of us who have experience with both procedures understand the processes and to the extent that, in any case, the reader understands the comparison made previously, we have a strong motivation for pursuing the concepts of concurrent new product development or concurrent project management.

As indicated in Figure 1.4, the preliminary business plan should be formulated early on. Every new product should be considered an incremental business, and the company should prepare a structured plan for the increment of business that the new product will bring in.

MOTIVATION FOR CONCURRENT NEW PRODUCT DEVELOPMENT

It is important to realize that new products must now be developed in one year or less, where before it seemed acceptable to allow two to five years. Global competition forces this situation. It follows that it is no longer acceptable for R & D to prototype a design and transfer this design to Operations. All functions within the company must work concurrently on the new product design and development.

It is important to realize that new products must now be developed in one year or less where before it was acceptable to allow two to five years. Global competition forces this situation.

It is not enough that the traditional product engineering, manufacturing engineering, and quality control functions attend design review meetings called by the research & development people. Even when the meetings are called regularly and there is good attendance, it is not enough. The functions downstream from R & D will usually attend, but since there is no assigned, in-depth responsibility at this early stage, they probably will not *really* participate in the design. To be effective, there must be active, productive participation with R & D by product, manufacturing, and quality engineering people and marketing, materials, and human resource principals, based on *all* functions having equal voice and veto power and based on in-depth assignment by top management. Key people must all be assigned and they must all exercise equal responsibility throughout the whole new product development process from the beginning of conceptual design through the development of the manufacturing processes and facilities. These activities should be structured around and based on concurrent engineering and on concurrent project management.

This exercise of responsibility must begin with the *heads* of Marketing, Engineering, Manufacturing, and Quality deciding concurrently the major issues of product identification, company goals, market objectives, functional specifications, broad product description, and recommended conceptual design. Final conceptual design must be the result of rich creativity by the team.

Finally, it must be stated that there is overwhelming evidence, from studying case after case, that the predominant companies successful in modern-day competition became triumphant only through the proper implementation of concurrent project management. Thus, success through implementation is the motivation.

As pointed out earlier, the input from marketing to the cross-functional team is a key factor. The marketing product manager maintains the focus on the customer. The cross-functional team must always keep the customer base in the loop. This is the essence of Quality Function Deployment. The focus on the customer is maintained throughout the whole product development process. Contact with the customer is maintained so that the team knows how to fine tune the specifications. This is especially important during the planning and scheduling phases.

For comparison, suppose marketing and manufacturing are not involved with R & D during the engineering design of the product. Suppose we are not careful and fall vulnerable to an inferior approach, which is quite common in many companies. The result is that the specifications continue to change too much for too long. An illustration is given in Figure 1.5, in which two different product definition graphs are shown. The older, more traditional approach, followed in many companies, is Graph Line I. The product specifications are not detailed well enough at the outset, with strong quality function deployment input from the customer base, and, secondly, manufacturing involvement is not deep enough at the outset of the new product development. Nor is marketing and manufacturing involvement deep and effective enough throughout the R & D phases. Consequently, the product specification continues to change throughout the R & D phases.

The result of a changing specification is that the development schedule is devastated.

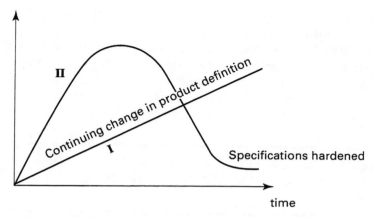

FIGURE 1.5 Change in product definition with time.

Graph Line II depicts the result of the preferred, concurrent project management approach. There is deep and effective involvement by all functions in this cross-functional teamwork situation. Product specifications are formed very dynamically in a rapid state of flux in the very early stages of planning. There is strong input from marketing at the outset of the program, along with manufacturing input to the design, including materials of construction and design-for-manufacturability requirements at the very beginning of the design. Consequently, the product specifications are detailed and hardened very early on.

The reason that Graph Line II does not go to zero, in Figure 1.5, as time increases, is that we want to leave ample room for creativity, even after the specifications are firmed up. We recognize that there will probably be some changes desired, and we do not want to stifle creativity.

Company 1, in the earlier comparison, probably follows Graph Line I; Company 2 probably follows Graph Line II.

THE PEOPLE INVOLVED

Companies successful in being highly competitive through concurrent management find it necessary to work on the people first and then the procedures for implementation. To handle normally existing but significant human behavioral issues, some successful companies find it extremely beneficial to work with a well-trained psychologist for some period of time during the team building phase. This may or may not be necessary in any given company. Suffice it to say that some companies find it quite advantageous and worthwhile.

People are, by and large, successful, and each person has achieved success in a manner that evolved over time. The person has established a routine, a procedure, which leads to success in terms of that person's goals. Having achieved certain goals, one records the manner which has led to such achievement. He or she is then reluctant to change.

It is important to study our procedures in light of the new demands placed upon us in this new business environment.

However, it is important to realize that our business environment has changed, and it is necessary that we study our set procedures in light of the new demands placed upon us in this new business environment.

The most important difference is that now global, highly competitive business conditions demand much faster product development. The procedures we followed before were adequate, but now we must change our procedures in order to develop products much faster. Instead of developing a product concept and a prototype in R & D and releasing it to Operations and Quality for them to develop the manufacturing methods, production facility, test specification, and test procedures all sequentially, it must all be done concurrently.

A key organization in any concurrent engineering implementation is the "customer," the person or group that receives the results of a project or task carried out.

RESPONSIBILITIES

The various functions listed in Figure 1.1 as those making up the cross-functional team are the following:

Marketing
Research & Development
Manufacturing Engineering
Quality Engineering
Materials Management
Service and Maintenance

Finance

Human resources

The responsibilities of most of the functions are well known. There are two that should be expounded on at this point. The industrial designer under R & D is responsible for infusing into the product the form, shape, and color of the final end product. These are very important elements because they are at the human interface level with the product. They provide the aesthetic appeal as well as the function.

The second function, which may not be as familiar to the reader as the others, is that of manufacturing engineering. In the case of product development, where the cross-functional team moves a product from concept to feasibility to development to commercialization, manufacturing engineering is a key discipline. Who are these people and what discipline do they infuse into the teamwork and into the product development?

Following is a list of responsibilities which characterize the profession of manufacturing engineering. The reader can readily perceive the importance of infusing this expertise into the process all the way from product concept to commercialization.

- Product design for manufacturability
- Producibility engineering
- Conceptualizing the manufacturing plan
- Manufacturing systems
- Manufacturing process design including materials processing and metrology
- CAD-CAM planning and participation
- Mechanization and automation planning
- Assembly and handling automation
- Economics and cost justification
- Cost estimating review
- The make or buy analysis
- Establishing work measurement standards
- Identifying capital equipment and facilities requirements
- Layout planning and material handling
- Planning for tooling procurement
- Manufacturing methods analysis and planning

- Flexible manufacturing or dedicated manufacturing analysis
- Computer integrated manufacturing analysis
- Machine and equipment utilization
- Scheduling
- Advanced manufacturing techniques

THE PSYCHOLOGY OF HUMAN BEHAVIOR IN CONCURRENT PROJECT MANAGEMENT

It was pointed out in the Introduction that most people are set in their ways. It is natural for us to resist change. However, in order to change from sequential new product development to concurrent new product development, it is essential to change the whole outlook and approach. Therefore, a change in human behavioral patterns is a major requirement. How to effect a change in organizational behavior to dissolve the traditional interface barriers between departments must be addressed.

Digital Equipment Corporation, one of the companies that has been quite successful in implementing concurrent engineering (CE), depends on a trained psychologist spending some quality time with CE team members. Xerox, Black & Decker, General Electric, and other companies, as well, address the human behavioral aspects of team-building. Certain employees of the company are assigned, full time, to team-building as a human resource responsibility. These people, in some cases not professional psychologists, are trained and are proficient in team-building and cross-functional teamwork concepts and implementation. These people hold workshops, form teams, and discuss the issues.

A change in human behavioral patterns is a major prerequisite to implementing concurrent project management.

In some cases, the training people hold workshops and form teams for practice only. In more advanced sessions, the trainers work with real cross-functional teams carrying out real product development. In each case, the team discusses and deals with human behavioral instincts and attitudes. Natural tendencies and resistance to change are discussed. The specialized psychological training and cooperation of the team results in the team members developing an *inner* desire to support each other throughout the life of the project. The team genuinely and sincerely experiences success when they feel that they cannot get along without each other, and the result is that they just naturally support each other.

Someone recently asked me, "But how much does it cost to carry out all the concurrent engineering planning and scheduling?" My response was, "Less than if you don't." If we just forge ahead and begin carrying out the project, we will suffer from not having planned and scheduled well. We will have to repeat many of the activities because of lack of good planning, and in the end, we will take longer to complete the project satisfactorily.

A change in the thought process must also take place at the upper management levels. Preferably, this change takes place first. My industrial students often tell me they have trouble selling their supervisors on the concepts of concurrent engineering. That is unfortunate for the students, to the extent that they believe in these concepts and want to implement them in their places of employment. The managers at the very top, though, do realize the importance of training from the top down.

TOP MANAGEMENT MUST CHANGE, FIRST

There must be top management involvement at the outset of each project. The first decisions regarding company goals and market objectives must be made by the people at the very top of each autonomous organization.

Unfortunately, it is the practice in many companies for the management in Manufacturing Engineering and Quality to delegate attendance at design review meetings at the earliest stages of new product development. This is wrong. It is at the outset in these early stages that strategic planning should take place with product concept in mind as well as company goals, marketing objectives, and total quality management. This high-level planning should be done by the top people in the principal functions. See Figures 1.3A and 1.4A. Later on in the development cycle, when the cross-functional team is developing the *correct* product to the *correct* functional specifications, based on sound market requirements, top managers can delegate responsibility. In addition to R & D, Product, Manufacturing, and Quality Engineering, the Marketing Product Manager must be an effective member of the cross-functional team. Also, the Service Department can infuse valuable design criteria.

Companies successful in concurrent project management have found that a whole new set of management rules must be put in place. Team-building and consensus management must be defined for the specific organization in each case and effectively implemented. Management direction in planning and implementing concurrent engineering must begin at the top, yet this is not always provided. I have found that too many top managers do not provide this direction. An even stronger statement can be made in the case of the traditional company: *It is recognized that top management has contributed to the traditional organization that now must be changed.* In the past, the president of the manufacturing company has appointed one individual to be in charge of R & D with the charter that "you be responsible for the design of the new product." The president has selected another individual to "be responsible for manufacturing including the development of the manufacturing processes," and yet another individual to "be responsible for the testability of the product." So top management, traditionally, has caused the organization to be somewhat *provincial* and, unfortunately, *fragmented.* Now top management along with all the other new product development participants must be "reprogrammed." All the people who are expected to be effective participants in implementing design-for-manufacturability, and the more all-encompassing concurrent engineering and concurrent project management, must undergo significant changes in their attitudes. These people must change, and it is against normal human characteristics to do this, for people resist change. But they must—and it has to begin at the top.

HOW TO GAIN TOP MANAGEMENT CONCURRENCE

To gain the attention and win the support of top management, it is necessary to lay before them the benefits in the form by which their performance is measured. Improved company profits is the key. Implementation of concurrent engineering requires changing behavioral patterns, training, and team-building. Time is required to do this. But just as it is correctly stated that quality engineering does not cost anything, neither does implementing concurrent project management, on balance. The benefits far outweigh any initial cost to achieve them. These benefits can and should be stated in terms of improved quality, reduced time to market, larger market share, and greater profits. General statements of expected benefits and abstract prospectives have no place here. The formulation and presentation should be specific and concrete. With one or more specific products in mind, a profit statement for each product with or without concurrent project management implementation can be formulated.

This can be done in terms of profit dollars on a timeline. It is interesting for top management to not only know the profit difference, with or without concurrent project management, but also to know *when* the profit will be realized. Successful companies do this with great attention to credibility. Careful attention to detail, conservative market estimates, which are usually somewhat effusive and illusive, and presentation of real case studies are favorably impressive to top management. The presentation must be so carefully prepared that it is highly credible and, of equal importance, *perceived* to have great credibility.

Figures 1.3 and 1.4 and the text describing them are pertinent here. It is meaningful to use charts such as versions of Figures 1.3 and 1.4 to illustrate the savings in money and in time to market by implementing concurrent project management. The figures should be tailored to exactly match the real product development project being considered. Such a careful analysis leads to a more accurate comparison in the specific case being presented as an example.

The benefits of concurrent project management can and should be stated in terms of improved quality, reduced time to market, larger market share, and greater profits.

As a somewhat different additional comparison, two plan networks can be prepared: one based on the traditional, sequential product development procedure, and one based on cross-functional teamwork, concurrent engineering. This preferred network is a subset of Figure 1.4 and, for reference, is found in Figure 1.4B. Resources and costs can also be detailed.

In Chapter 3, two estimates are derived from the carefully generated plan network. Both are based on the estimates of project completion being a Gaussian (normal) probability distribution. This can be proven valid. The normal probability distribution leads to an expected project completion date, which is mathematically correct. However, it must be remembered that the normal distribution implies that there is only a 50 percent probability of completion by this date. A much more interesting date is the one with 90 or 95 percent probability of success. This, too, is mathematically derived and therefore has great credibility. People at all levels of management are interested in these estimates with regard to how they relate to their own respective organizations.

COMPANY-WIDE CONCURRENCE

Following the buy-in by top management will be the gaining of concurrence by R & D, Marketing, Manufacturing, and Finance to inwardly and outwardly want to pursue and properly implement concurrent engineering. The initial objections are usually characterized by people stating that they desire concurrent engineering but who do not really follow their statement. One may hear Manufacturing and Quality objecting to assisting in the development of the new product because of the following: transcending, current production problems on the floor that prevent this month's shipments; limited budget and personnel to participate in new product development; and lack of training to understand the new product technology. These, if they do occur, are all short-sighted excuses. Current production problems are probably caused by not having properly implemented concurrent project management in the past when the new, troublesome product was "developed." Limited budget is not a valid excuse because in the end concurrent management costs less. The new product technology must be learned by all at some point. It does not cost more to learn it earlier. Transcending all of this hyperbole is that to survive as a competitor, the manufacturing company must shorten the time to market, hold market share, and realize a significant profit. To do this, concurrent project management *must* be implemented! The psychological training covered in the previous section is important in this process.

When R & D, Marketing, Manufacturing, and Finance prepare a product/profit plan based on concurrent project management, all in concurrence from a well-established team, top management will then also concur. Having the whole organization concur—including top management as well as the working team—will bring the desired results.

PRODUCT CHAMPION

Much attention has been paid by organizations to improving the transfer of technology. Many people have attended seminars, workshops, and short courses to study transfer of technology techniques. Within a company, the strongest, most effective transfer of technology can be achieved by the principals (between whom the transfer must take place) by performance within a cross-functional team. Because they are all performing their individual responsibilities as members of a cross-functional team, the technology is inherently understood by all, and it is understood concurrently. It is not, then, a matter of transferring technology from R & D to Manufacturing Engineering at some point after prototype development. Technology transfer should be an integrated, continuing process taking place within the team from the time the team first begins working together in the early conceptual design stage of new product development.

> *It has been proven that the greatest success has been achieved where a project leader is selected early on and where this individual is an effective product champion.*

In carrying out concurrent engineering, it has been proven that greatest success has been achieved where a project leader is selected early on and where this individual is an effective product champion. A product champion is one who lives and breathes the product as it is developed and "champions" its cause throughout the company. The champion effectively drives the development through any obstacles that arise. The champion, by definition, is effective.

Several prerequisites qualify a Project Leader/Product Champion. This individual must have good product knowledge, be a hard worker, and have good interpersonal skills. Where a manager has failed, the manager was usually found to be lacking in one or more of these areas. A broader, more detailed list is given in Table 1.1, "Characteristics of a Leader."

The Project Leader and the remainder of the team study the technical feasibility, write a marketing plan, develop the design, study resources, and construct company-wide prototypes to test both the technical feasibility and the manufacturing processes. These activities are carried out by the cross-functional team all performing product design, quality, and manufacturing process development concurrently. The Project Leader/Product Champion is an extremely important element in project success.

CASE STUDIES—A REFERENCE TO CHAPTER 9

There is strong motivation for a company to properly implement concurrent project management. The best way to convince the reader of this is to present several real case studies. There is nothing like actual case histories to make a convincing argument in favor of a proposed discipline.

TABLE 1.1 Characteristics of a leader.

A principled person

Absolute integrity

Levels with people

Good communicator

Knowledgeable in chosen discipline, good at training

Well informed with regard to policies, fringe benefits, customers' needs — good at imparting this knowledge

Always returns phone calls

Asks for advice—follows good advice

Good listener

Gets facts, then decides

Clean-cut and charismatic—has the appearance of a leader

Has confidence—instills confidence

Creative

Forges new paths

Willing to fail and fail, again, in order to finally succeed

Willing to let others fail

Must be a good loser

Takes individual responsibility

Hard worker

Fulfills own needs

Fulfills needs of others

Good at marketing employee's services

As time elapses, these case histories will become dated. However, the compelling argument for cross-functional teamwork in concurrent project management will become stronger and stronger as the economy becomes more globally integrated, causing stiffer and stiffer competition. Therefore, the discipline advanced by a study of these case histories will remain ever important.

Four case studies will be presented in Chapter 9. All are new product development cases that are directly related to concurrent engineering. Three are success stories; one was not so successful and was the result of improper management and is presented only for comparison. One of the success stories is taken from a conference attended in 1989. The other three are direct experiences from the author's places of employment.

IMPLEMENTATION OF CONCURRENT PROJECT MANAGEMENT

In line with the concepts presented so far, it is necessary to stress implementation issues and criteria. Some that are particularly relevant to this discussion follow:

(a) Design engineers should receive some training in manufacturing, quality, and service.

(b) Everyone on the design team should be given equal rank.

(c) The design and development of the manufacturing processes and quality test procedures, methods, and overall quality engineering are as important as the design of the product.

(d) The team must continuously focus on the customer requirements.

(e) The team must continuously focus on the manufacturing requirements.

A more comprehensive list of the issues, which this author has named the Design-It-Once criteria, is presented in Table 1.2.

Now, especially after studying Table 1.2, it seems that there are some problems raised by the individual company employee who is being asked to perform as a member of a team. Management, responsible for morale, must address these problems. How is the individual's salary increase determined by team performance? How do you handle ego when the individual wants his or her name on a product when the product is developed by a team? How do you motivate people, having naturally different interests, to perform as a team? The answer to the first question is that, even as a team member, the individual has unique expertise that the team is counting on. Therefore, how well this person carries out responsibilities, while a member of the team, determines fair compensation. Also, when the team becomes successful in developing a highly competitive product, the team will be rewarded and, by definition, the individual is part of the team and so will realize the reward. This is quite satisfying to one's ego. In any case, benefits to the company, the larger body of people, always transcend the singular benefits to a given individual.

The study then progresses to one of how to properly implement concurrent engineering (CE). In many cases it is heard expressed that it is well known that concurrent engineering is good, but how is it carried out — how is it implemented? Individuals know that CE is important; however, what many lack and strongly desire is knowledge in how to realize benefits from it. What the individual is expressing is that it would be

TABLE 1.2 New product cross-functional development team issues and criteria.

1. Team size limited to six or eight people.
2. Include people at the outset and all along the way who can prevent engineering changes later.
3. Properly integrate these people—get the team members so integrated that they cannot get along without each other.
4. Must enable each individual to carry out his or her own responsibility.
5. Must give each individual team member the required tools.
6. Give them all a common database.
7. The same people who specify the design are those who must be there at the end to make the product work.
8. Everyone should go beyond understanding the written specifications to an understanding of *all* of the customers' needs.
9. The new product must do everything that the competition's product does that is of value to the customer.
10. Must train everyone in QUALITY, COST, and SCHEDULE —there is usually need for improvement in all three areas.
11. Involve hourly employees early in the development process.
12. Involve the service department early on.
13. Must make sure that everyone does not gyrate to an R & D function.
14. Assign the design engineers to Manufacturing and Quality Assurance for long enough for them to "feel" the requirements of these downstream functions.

highly beneficial to know how to effectively implement concurrent engineering, bearing in mind that, in each case, the individual is performing within a given organization and wishes to strike an optimum balance between individual pride in one's work and individual recognition on the one hand, and good teamwork results on the other. The organization must change and also the individual must know how to effect the required changes within himself or herself for the overall good of the company. As important as the team is, the larger body of people composing the company is even more important.

Some companies have successfully implemented concurrent engineering—many more have not. Many have addressed the need and yet have failed to properly implement it. Unfortunately, a great many companies still adhere to the traditional, sequential procedure of developing a theory and a prototype in R & D, then passing the prototype to Product Engineering and, in some cases, to Manufacturing Engineering for facilities and process development. Then, later, Quality Control attempts to qualify units for shipment after production.

The traditional, sequential procedure is no longer competitive. The preferred discipline of concurrent project management detailed herein is mandatory.

CREATIVITY

The selection of the best possible product design concept has such a far-reaching effect that we must study this matter and discuss it adequately to insure proper selection. The cross-functional team must study, repeatedly, all conceivable product design concepts to identify the one that leads to the minimum number of component parts, is the simplest to manufacture, has excellent quality in terms of both useability and reliability, is the least costly, is the easiest to insert into the larger system in which it will be used, and is the easiest to service.

In order for the team to select the best concept, it must have all possibilities in mind. Where do all these ideas come from? How does the team make sure it has a full bag of possibilities to choose from?

The team operates from experience. It knows of all similar products both within the company and among all the related competition. It knows all the science embodied in the prior art and the science underlying the new product. It knows the methods available. And, yet, it will prove beneficial for the company to carry out additional brainstorming on the new product. By definition, the project is that of developing a new product. It has never been developed before in exactly the same form and function. To be sure, all the experience in the team will be required. However, because the project calls for developing a new product, brainstorming will usually turn up something new that will lead to a better product concept.

BRAINSTORMING

Principles of good brainstorming are the following:

1. Gather people from within the project and one or two from outside the project to participate. A person or two from another project, group, building, or facility who are cleared to discuss the project can be very productive because a fresh view is taken by each.
2. Distribute a written problem statement and notice of the meeting.
3. Meet in the morning. People are more creative then.
4. Meet someplace for three hours where secretaries, supervisors, and employees cannot contact the brainstorming participants.

5. Make sure everyone understands the problem to be discussed.

6. Disregard limitations such as cost and schedule, for now.

7. Do not allow any criticism of anyone's ideas —not even constructive criticism.

8. Make sure each and every idea is recorded—have a separate recording secretary so that brainstorming participants are not encumbered by having to take notes.

9. Allow only one person to speak at a time.

10. Gather all ideas, select the one or more that are the best, and move forward with the project.

The following illustration involving a physics student illustrates creativity and points out the advantages of proper selection.

CREATIVITY

The process of creativity is a mysterious and interesting one. It is brilliantly described in Alexandra Calandra's *The Teaching of Elementary Science and Mathematics.*

A student refused to parrot back what he had been taught in class. When the student protested, Calandra was asked to act as arbiter between the student and his professor.

Calandra stated: I went to my colleague's office and read the examination question: "Show how it is possible to determine the height of a tall building with the aid of a barometer."

The student had answered: "Take the barometer to the top of the building, attach a rope to it, lower the barometer to the street, and bring it up, measuring the length of the rope. The length of the rope is the height of the building."

Calandra stated: A high grade is supposed to certify competence in physics, but the answer did not confirm this. I suggested that the student have another try at answering the question. I gave the student six minutes, with the warning that his answer should show some knowledge of physics. In the next minute he dashed off his answer, which read: "Take the barometer to the top of the building and lean over the edge of the roof. Drop the barometer, timing its fall with a stopwatch. Then, using the formula $S = 1/2\ at^2$, calculate the height of the building.

Calandra said: At this point, I asked my colleague if he would give up. He conceded, and I gave the student almost full credit.

In leaving my colleague's office, I recalled that the student had said he had other answers to the problem, so I asked him what they were.

"Oh, yes," said the student. "There are many ways of getting the height of a tall building with the aid of a barometer. For example, you could take the barometer out on a sunny day and measure the height of the barometer, the length of its shadow, and the length of the shadow of the building, and by the use of a simple proportion, determine the height of the building."

Calandra continued to recite the incident: "Fine," said the student. "Take the barometer and begin to walk up the stairs. As you climb the stairs, you mark off the length of the barometer along the wall. You then count the number of marks, and this will give you the height of the building in barometer units. A very direct method."

"Finally," he concluded, "there are many other ways of solving the problem. Probably the best," he said, "is to take the barometer to the basement and knock on the superintendent's door. When the superintendent answers, you speak to him as follows: 'Mr. Superintendent, here I have a fine barometer. If you will tell me the height of this building, I will give you this barometer.'"

It can be seen from the above illustration that the best way of carrying out the activity of determining the height of the building was for the person responsible to approach the building superintendent and merely ask for it. The result was the most accurate, the time required to carry out the activity was the least, and the cost of carrying out the activity was lowest, depending, of course, on the cost of the barometer and the accuracy required in the answer.

This kind of creativity should always be exercised in project management. In a great many cases the most creative and productive solution will come from the good people within the company. In some cases, the best solution will be to go outside the company for the answer. The parallel in the physics student illustration was to go to the superintendent instead of limiting the choices to those the student could provide by himself, or limiting the choices to the one someone else wants, which in the illustration was that of using the barometer as an altimeter.

OVERALL PROCEDURE

The *Overall Procedure* for implementing concurrent project management is given next. We will follow this procedure in presenting several examples in succeeding chapters.

The *Overall Procedure* is explained at each element in its presentation. In addition, as this procedure is carried out in the examples, each element is explained in detail. Individual examples of each element are presented.

OVERALL PROCEDURE FOR CONCURRENT PROJECT MANAGEMENT

1. Clarify the **FUNCTIONAL SPECIFICATIONS.** Study the elements of planning to make sure you and your teammates have properly addressed all aspects of planning. Clarify the company objectives and the marketing objectives of the new product development. Study competition. Determine customers' needs. Write product description. Write final functional specifications. Document agreement with the management team by everyone on the team signing the final specifications.

2. Detail the product specifications. Identify and establish the cross-functional team. Study competition. Study the functional specification. Select the design concept. Detail all aspects of the physical product. Rely on experts. Write a **DETAILED PRODUCT DEFINITION.** Document agreement within the cross-functional team by having all members of the team sign the detailed product definition.

3. Describe the project. Elements 1 and 2 above, describe the product. This element, the **PROJECT DESCRIPTION**, describes the project of developing the product. List all the tasks constituting the project—list approximately forty or fifty tasks. Rely on experts. Write a detailed project description. Document agreement within the team in the same manner as above.

4. Generate the **PLAN NETWORK.** List all the tasks/activities that compose your project. Prioritize them. Indicate after each activity the one or more other activities that immediately precede the one under study. Draw a rough network to illustrate the correct configuration. Rely on experts. Study resources. All members of the cross-functional team must sign the plan network.

5. Validate the **RESOURCE ALLOCATION.** Draw the work breakdown structure and assign the different talents to corresponding tasks.

Obtain approval from the source managers for scheduled allocations. Modify the plan network configuration if necessary. Redraw the plan network on a linear time scale. Identify the critical path. All members of the cross-functional team must sign the revised plan network.

Draw the resource allocation in the form of a histogram/vertical bar chart on the same time scale as used for the plan network. All members of the cross-functional team must sign the resource chart.

6. Generate the **COST SCHEDULE**. Draw a cost curve as a function of time using the same time scale as used for the plan network and the resource histogram. All members of the cross-functional team must sign the cost curve.

7. **REPORT** the plan to management and request funding. Because experts were relied upon for the above planning, the best possible plan has been put in place. The project leader has the responsibility of presenting the plan for approval.

This procedure is repeated at certain milestones in this book; an increasing number of leading paragraphs will be in **boldface** type, as we move through the process, to indicate the extent to which the procedure has been covered in the text up to that point.

Also, a detailed version of the procedure is given on pages 44–46.

TOTAL QUALITY MANAGEMENT

Total quality means quality of performance, of leadership, of service, of product, of company/employee relationship, at all levels, in all departments, and in every function.

Total quality management (TQM) is a term that pervades many organizations, and TQM discussions abound. What matters, though, is implementation. How do we deliver better quality to the customer? Here, *customer* means that person or organization downstream that receives the output from another person or group. The output may be a drawing, report, product, or a service. With this definition, everyone within a company has a "customer." We are all responsible for delivering total quality from our office or station to the "customer."

Joseph Bellefeuille stated in the *IEEE Engineering Management Newsletter* of July 1993, "Total Quality Management is an interlocking arrangement of procedures and practices which ensures that ALL

EMPLOYEES in every department are adequately trained and directed to CONTINUOUSLY IMPLEMENT CONGRUENT IMPROVEMENTS IN QUALITY, SERVICE AND TOTAL COST such that CUSTOMER EXPECTATIONS are met or exceeded."

This book provides an *Overall Procedure* for implementing total quality management in terms of concurrent project management.

PROBLEMS

1.1 What are the various functions within a manufacturing company that typically compose the cross-functional team?

1.2 Why is it important for the people representing various functions to work together as a team, in terms of quality of product, time to market, development cost, and product cost?

1.3 What is the relationship, by definition, between design-for-manufacturability, concurrent engineering, and concurrent project management?

1.4 Define the term *concurrent engineering*.

1.5 Define the term *concurrent project management*.

1.6 What are the *two* meanings of the word "concurrent" as used in this book?

1.7 What are the four business combinations in which concurrent project management can be implemented?

1.8 What is the primary difference in the two new product development procedures as illustrated in Figures 1.3 and 1.4, in terms of who provides the conceptual design and the prototype?

1.9 Who carries out the product design in the traditional process?

1.10 Who carries out the product design in the preferred process?

1.11 In the preferred process, what functions are carried out concurrently?

1.12 State, in twenty-five words or fewer, the motivation for concurrent project management.

1.13 Why is it a prerequisite, in preparing for concurrent project management, for the people to change their mind-sets?

1.14 What is the best way of gaining top management concurrence in concurrent project management?

1.15 Define the term *product champion.*

1.16 List the five characteristics of a leader from Table 1.1, in your order of priority, that are most meaningful to you.

1.17 List the five issues and criteria from Table 1.2, in your order of priority, that are most meaningful to you.

1.18 Creativity is an important characteristic in the new product development team. Why is it essential to implement creativity?

1.19 List the seven elements, in two or three words each, in the overall procedure for concurrent project management.

2 CONCURRENT PLANNING

INTRODUCTION

This book teaches technical project management based on the modern discipline of concurrent engineering, or, more broadly, concurrent management. That is, it is based on team-building and teamwork across all the required functional disciplines in an organization developing, producing, and marketing products.

Although there are many aspects to the process of concurrent product management covered herein, it is still the case that planning, scheduling, and controlling are paramount, and they will all be presented and treated in great detail. Without proper planning, no organization can long survive. The best way to schedule a project is by way of preparing a sound, detailed, meaningful, practical plan as a prerequisite. And certainly the best way of controlling a project is through careful planning at the outset. Imagine the opposite. Imagine that the project is poorly planned so that whatever plan is made is unsound, not detailed, not meaningful, and/or impractical. Try delegating this plan to the product development team for execution and control. It would be impossible to control this project so that performance, cost, and schedule are well managed. So, as a test, it is sometimes just and prudent to imagine the opposite to the preferred process. It helps to put these matters in their proper perspective as it establishes the proper frame of reference. Then, we can move forward with strong, directed motivation. Therefore, let us agree that the team should prepare an excellent plan. The point to be made is that once we have a sound plan and we have based a realistic schedule on this plan, then it will be relatively easy to control the project.

Let us turn then to planning.

Concurrent planning is the planning of all the activities and tasks composing the overall technical product development project. In particular, concurrent planning is generating the plan by working together as a cross-functional team.

All the activities should be studied by the team. It is as important for the correct members of the team to have early, continuous, and effective inputs on manufacturing engineering and quality engineering as it is for the research, design, and development engineering personnel to have inputs into the new product design.

This planning of the manufacturing processes concurrently with the design of the product by teamwork, and gaining concurrence of *everyone* on the team, is the essence of concurrent engineering. It is important for the design and development of the quality engineering and the design and development of the manufacturing processes to be carried out concurrently with the design and development of the product. In fact, input from the service and maintenance personnel should also be infused into the design.

It is important for the design and development of the quality engineering and of the manufacturing processes to be carried out concurrently with the design and development of the product.

In Chapter 1, we defined concurrent engineering and illustrated its relationship to design-for-manufacturability and concurrent project management. Concurrent project management embraces all the various functions that can best work together to carry out the product development project. These functions include not only R & D and manufacturing but also quality engineering, marketing, product management, materials management, human resources, finance, and service. The R & D function must include industrial design that infuses form, shape, color, and ergonomics. Inputs from these functions provide a design that is complete, and preclude having to redesign the product.

To move forward now with the planning process, we will define, in this chapter and the next, seven basic elements composing the project plan. They are the following:

1. the Functional Specification
2. the Detailed Product Specification
3. the Project Description
4. the Plan Network
5. the Resource Allocation
6. the Cost Schedule and
7. the Report/Proposal.

Each element has a very special purpose. Each is required. If one is missing, there is significant degradation of the plan. Each fits very nicely into the plan and is complementary to the other six. The above characteristics of the elements of the plan are true only if each is of high quality. It is the purpose of this chapter to give the reader a complete understanding of the first three elements:

1. the Functional Specification
2. the Detailed Product Specification and
3. the Project Description.

In addition, the plan network will be partially generated. We will present a rough configuration. However, for the network to be complete, we must develop it in terms of scheduled activities. Scheduling will be covered in Chapter 3. The remaining three elements, which are the resource allocation, cost schedule, and report/funding request, must be completely defined in later chapters.

The reader should refer to Figure 2.1, which illustrates the process that is followed in bringing all the important elements of the *Overall Procedure* into agreement. The process of refining the accuracy and utility of the various major elements of planning and scheduling are extremely important.

1. Top management provides the Functional Specification.

2. (a) The next level of supervision, reporting to top management, studies the Functional Specification and writes a Product Specification. This further defines the product and enables identification of a full cross-functional team.

(b) A cross-functional team, which is an expansion of the next level of supervision, of Step 2a above, is formed. This team studies and details, even further, the Product Specification.

At this point in the process, there will probably be some feedback to management. Some facts are now known about the product that may influence all the people involved so far to change the Functional Specifications somewhat. For example, in detailing the Product Specification, it may be determined that the development time stated in the Functional Specification was too short.

There will be one or two iterations in this process until the Functional Specifications and the Detailed Product Specifications are brought into agreement.

3. The cross-functional team writes the Project Description.

4. The team generates the Plan Network.

5. The team generates the Resource Allocation. It is usually found that resource limitations cause some amount of reconfiguring of the Plan Network. Some parallel activities may have to be converted to sequential activities, for example. Therefore, there are one or two iterations back through Step 4 until the network and available resources agree.

Also, since this schedule is now known more accurately, the one stated in the Functional Specifications may have to be modified.

6. (a) The cross-functional team generates the development cost estimate. This estimate is based on a hardened schedule of activities and corresponding resources. Therefore, it has great credibility.

(b) The cost stated in the Functional Specification may have to be changed at this time.

Once the above process is fully carried out, we will have all the elements of the *Overall Procedure* in strict agreement, where, when they were first provided, they were not. We can now write the Report.

7. The Project Leader can now write the Report which is a summary of the bottom line results of bringing all the elements of planning and scheduling into concurrence. The Report is actually a request for funding the project. We can, therefore, also think of the Report as a Proposal.

Top management provides **Functional Specification**

Next level of supervision writes Product Specification

Cross-Functional (CF) team identified

Cross-functional team writes **Detailed Product Specification**

Team discusses any technical differences with top management and modifies Functional Specification, if necessary

C-F team writes **Project Description**

C-F team generates preliminary **Plan Network**

Team generates **Resource Allocation**

C-F team revisits the **Plan Network** to make it conform to available resources

C-F team generates development **Cost Schedule**

Team discusses Cost Schedule with management to bring everyone into concurrence

Project Leader submits **Report/Proposal**

FIGURE 2.1 Flowchart illustration of concurrence.

FUNCTIONAL SPECIFICATION

The functional specification is a description of the product to be developed as it is viewed from the top level down. It usually comes to the product development team from a higher supervisory level. It describes the functions the product is to provide. Compared with the detailed product specification, the functional specification is relatively short. It may be only one-half page or one page long; the detailed product specification may be four pages, ten pages, or even longer. The lengths of these two depend, certainly, on the specific product under development. The point is that they serve two different purposes. The functional specification points out the functions and features to be provided. It is relatively brief compared with the detailed product specification, yet it is complete in providing its purpose. It originates from higher-level management people who have the primary responsibility for deciding company goals and marketing objectives. It presents the broad characteristics of the product as it will be viewed by the customer base in the future. It is purposefully general, for top management delegates the responsibility for detail to the cross-functional team.

The functional specification presents the broad characteristics of the product as it will be viewed by the customer base in the future.

The detailed product specification, on the other hand, details the electrical, electronic, mechanical, hydraulic, pneumatic, chemical, thermal, and/or aeronautical hardware and software characteristics; the external physical characteristics; the means of interfacing with other products or systems feeding into it or accepting signals from it; and the size and other dimensions needed to facilitate installation of the product. Therefore, the detailed product specification may be five or ten times as long as the functional specification. I have seen detailed product specifications 150 pages long.

The functional specifications are usually extremely brief compared with the detailed product specifications.

The functional specification must come first. It is extremely important as it defines the product. It derives from top-level considerations regarding the product line the company wishes to promote and market. In terms of company success, the proper definition of the product compared with already existing products in the company and in the competitive marketplace is all-important. We suggest that more and more attention be paid to what the future user population desires in terms of product function. The functional specification then derives from these customer needs. Many people call this discipline quality function deployment.

The functional specifications are very brief compared with the detailed product specifications.

Also, the functional specification contains elements descriptive of the project regarding schedule. It should, at least, specify the desired time of availability of the product. It should also specify not only the proposed cost of the product but also suggest a limit to the cost of the project. Following is a list of the elements of a functional specification.

Elements of a Functional Specification

- Assignment statement
- Type of product
- Several top-level functional features
- Name(s) of competitive products
- Expected cost of product
- Expected cost of development
- Expected development time
- Statement that it is expected that there will be (or will not be) changes in manufacturing facilities required
- Places for signatures with dates (to be signed and dated after time and cost have been scheduled—see *Overall Procedure)*
- The functional specification should be from a specific individual of authority, and it should be addressed to a specific individual with responsibility.

An example of a functional specification is given in Figure 2.2.

Top-level management expects the cross-functional team to study the functional specification, and to buy-in and to sign-in, indicating general agreement at some point in time. Quite often the team will want to change the functional specification prior to buying-in. It may be that the stated, expected function, schedule, or the cost cannot possibly be met. Top-level management will expect and should demand this kind of feedback to insure correctness and concurrence with the functional specification. So it is a top-down assignment with bottom-up feedback dynamics, with a focus on customer requirements until management and the team are all in agreement. Coming to early agreement is extremely important. This a prerequisite to successful management of technology.

There is a new product development process philosophy that should be explained here. The previous paragraph implies crispness in the process, and we should strive for such a process. However, we must also allow flexibility.

There must be agreement that the specifications can be changed. The team should comment almost immediately on the veracity of the specification. Also, it is, in most cases, prudent not to sign-in until after the cost schedule has been formulated. The cost of the development project depends on the specifications, both functional and detailed, and since the cost specification is a part of the functional specification, then all factors

To:

FUNCTIONAL SPECIFICATIONS
for a
HOME

Design and develop a home. Generate and provide a detailed design of the house, including all blueprints and specifications regarding materials of construction and workmanship.

The house will be a southern colonial type attractively located on 21,780 square feet of land located at 24 Sweetbriar Drive in Cranston, Rhode Island. The type of construction will be standard joist and stud construction as opposed to post and beam construction.

There will be 3,000 square feet of living space, and, in addition, there will be a deck and walk-out basement built to the architect's specifications.

Provide an attractive landscaping and overall architectural appearance.

All building codes and zoning laws and restrictions must be met.

Labor cost must be in the range from $35,000 to $45,000. The total cost must be in the range from $100,000 to $125,000.

The home must be completed, including all landscaping, and must pass inspection to all specifications in preparation for the official closing on or before (specify date).

	Name	Date
Project Leader:	_____	_____
	_____	_____
	_____	_____
	_____	_____

(Do not sign until after time and cost schedules are established)

From:

FIGURE 2.2 Functional specifications for a house.

must be in agreement. That is, the functional specification, the detailed product specification, the project description, the plan network, resource allocation, and cost schedule must all be in agreement. It is best to accept the functional specifications for a guide, but with an understanding that, in the process, schedule and cost will depend on the details to be formulated later. The point is that at some time, as early as possible, the cross-functional team will sign-in, validating the functional specifications as a document.

Reference is again made to the process illustrated in Figure 2.1.

True schedule and cost will depend on the details to be formulated later.

The underlying reason for this process is to establish and maintain good communications. It is important for everyone on the team to have the same information and database.

We must remember that all of this planning and scheduling is taking place before the development of the product begins. Once sufficient study by the cross-functional team has been carried out to validate the time and cost schedules, then everyone can buy-in to the specifications. It is during this process that the specifications are often changed.

The argument is often presented that the time and cost schedules cannot be hardened until after the specifications are hardened. This is true. The point, here, is that the process is a recursive one; it is a successive approximation and the specifications, the time schedule, and the cost schedule converge asymptotically until these elements are all in agreement and the members of the cross-functional team are all in agreement.

Above all, this convergence, this agreement between specifications and schedules, and concurrence by all, must take place prior to the beginning of the product development. Many new product developments fail to meet satisfactory company-wide goals because early agreement/concurrence is not achieved.

The functional specification is really the starting point in the proposed new development as determined by top-level management. It is to be interpreted as an assignment to those responsible for teamwork development of the product.

An example of a functional specification of an encoder is given in Figure 2.3.

To:

FUNCTIONAL SPECIFICATIONS
for an
INCREMENTAL OPTICAL ROTARY ENCODER

Design and develop an encoder. Generate and provide a detailed design of the encoder including all detail and assembly drawings and bills of materials.

The encoder will be an incremental optical rotary encoder. It is desired to supply to the market a new, heavy duty, flange or servo-mounted encoder for industrial use. It should be no more than 3" in diameter and 3" long. It should have long life components, including the light sources. The speed of response should be at least 100 khz. The design will be in accordance with all detailed product specifications to be generated as the second step in the specifications process.

The cost of the product is to be less than $350.

The cost of the design and development project should be less than $85,000.

This new product should be available for the market in ten months.

Study competitive products.

It is expected that existing manufacturing facilities will suffice.

	Name	Date
Project Leader:		

(Do not sign until after time and cost schedules are established.)

From:

FIGURE 2.3 Functional specifications for an incremental optical rotary eccoder.

Finally, all subsequent documents such as the detailed product specification, project description, plan network, and so forth, must agree with the functional specification. It is for this reason that the functional specification must be meaningful, practical, and realistic, and must describe a product that can actually be successfully developed. It cannot contain elements such as unrealistic cost and time schedules that are highly improbable and even impossible. Therefore, let us move on to the detailed product specification and then return to the functional specification, later, for final editing approval.

THE DETAILED PRODUCT SPECIFICATION

The detailed product specifications are extremely important. They completely describe the end product. Product specifications for a house would be the blueprints and the written detail regarding all building codes, materials of construction, interior and exterior finishing, and every additional detail, completely defining the house and distinguishing it from other houses. Product specifications for a home would include not only those for the house but also the complete definition of landscaping, curb design, garage, driveway, location with respect to adjoining properties, and all the other issues important to the homeowner.

Often a new product is an upgrade from an existing product. Sometimes the mistake is made of specifying only the differences from the older product. The mistake is that the base product, forming the basis of the new design, is often deficient in some respects. It is better to treat the upgrade as a whole new design and write the specifications accordingly. Then nothing gets lost in the transition. Take the example of the company wishing to add a remote readout instrument to a primary product such as a flow meter. The flow meter had previously been read at the meter and had an accuracy of ±2 percent. The new specifications were written as having an overall system accuracy of ±1 percent. If insufficient care were taken, the upgrade development team might develop the remote readout and overlook the accuracy deficiency in the primary meter. It is better to treat the new product as an overall system instead of just the new part. This prevents misinterpretation and trouble later.

The detailed product specifications range from three or four pages to 100 pages or more, depending on the specific product.

The detailed product specifications may be three or four pages, ranging to 100 or more for relatively complex new products. The author recently viewed specifications for a new building that were 150 pages long. Specifications for a custom chip were eighteen pages long.

The detailed product specifications must conform to the functional specifications.

The detailed product specifications must conform to the functional specifications. There must be no disagreement between them. Although they serve two different purposes, they must be in complete agreement.

Sometimes, as the product specifications are detailed, it becomes evident that the functional specifications must be modified. For example, suppose the functional specifications state that the cost of product should be no more than $125,000. Suppose, further, that we find as we detail the specifications that the *realistic* cost is $135,000. The marketing and business plan must be checked. If it is found that we can stand $135,000, then we will change the functional specifications. If the market cannot stand such a high price, then the specifications must be restrained to a more competitive level. This is a critical decision.

The test as to whether there is enough detail in the product specifications is whether the following question can be answered affirmatively: Is it possible that any future delay in time to market could be traced back to inadequate detail and/or validity in the detailed product specifications prepared at the outset of the development program? Could there be unnecessary redesign and frustration in the cross-functional team caused by misunderstanding of the details of the definition of the product to be developed? If so, then more detail must be added to the product specifications and the specifications must be clear to everyone before the development begins. Moreover, all members of the team must concur regarding the final, detailed content of the specification.

The cross-functional team should study competitive products in order to properly specify the product to be developed.

Sometimes, it is even advisable to spell out in the product specification the features that similar products in the marketplace have that this new product will not have. This often prevents misunderstandings between

Engineering and Marketing, for example. This can happen even though both departments are fully represented in the same cross-functional team.

Consideration of competitive product features broadens our view. It may be decided to include one or more features a competitor has. It may be decided not to. In either case, the decision regarding the overall product content is a more informed one.

A great deal of creativity and diligence must be applied in order to generate a detailed, comprehensive, complete, and accurate specification. The correct mix of aggressiveness and patience must exist. The team may be inclined to "get on with it," feeling that they have reached the point of diminishing returns, when, in fact, they have not. Secondly, sometimes even though the team's intent is good, they just cannot think of all the details. It helps to study specifications for other similar products in the company's product line and also those for competitive products.

Another suggestion is to write, in rough draft form, at the very outset, the operator's manual, the service manual, and any other such documentation that will accompany the product. This activity will serve to remind everyone of some of the important details that are usually otherwise forgotten.

An example of a detailed product specification is given on the next four pages, with permission from Dynamics Research Corporation.

HARDENING THE SPECIFICATIONS

Following are issues and questions and answers regarding how to harden specifications:

1. What are the broad guidelines of what we are to do? This was answered once, but now there is a cross-functional team. So, now we can refine these guidelines.

2. What are the functional specifications? As a team, we should review the functional specifications, and maintain a focus on them.

3. The detailed product specifications must be derived from the functional specifications. They will change somewhat as the project is carried out, but, when all is said and done, these two different types of specifications must agree.

4. What exactly are we to provide, as a product?

   Consider a product based on a custom integrated circuit chip. The chip supplier and the user company agree on $115,000, say, for the supplier to develop the chip to the company's specifications.

MODEL 25

Incremental Optical Rotary Encoder

- **2.5" diameter, heavy duty encoder**
- **LED light source**
- **Flange or servo mount configuration**
- **Sealed bearings are standard for maximum reliability**
- **100KHz count channel frequency response**
- **50KHz zero reference frequency response**
- **Optional 2X, 4X, or 5X cycle interpolation electronics**

The Model 25 has been designed for rugged, OEM applications where reliability is a prime consideration. Features such as 100K hours MTBF, LED light source, 40 pound axial/35 pound radial shaft loading capability, 100KHz (count channel) and 50KHz (zero reference) frequency response, and 50G's shock specification allow the Model 25 to be used in machine tool, robotic, and other harsh environment's rough usage applications. The Model 25's 2.5" diameter housing is available in flange or servo mount configurations, and can be provided with a variety of bolt circles allowing easy mounting. For high ambient noise or long transmission length applications, the Model 25 can be supplied with 8830 or 88C30 line drivers. In line driver, (or cycle interpolation) configurations, the Model 25 provides a ¼ cycle wide gated zero reference output designated as GZ. The cycle interpolation option allows a maximum resolution of 60,000 counts per shaft revolution.

Product
Data
Sheets

SPECIFICATIONS

ELECTRICAL

Resolution range:
- Discs can be provided with up to 3,000 lines, providing a maximum of 3,000 cycles per shaft revolution exclusive of either internal or external cycle interpolation. (To 12,000 counts per revolution with external 4X circuitry only, or to 60,000 counts per revolution with internal 5X and external 4X circuitry.)

Light source:
- Gallium aluminum arsenide LED rated for 100,000 hours MTBF (manufacturer's specification).

Light sensors:
- Photovoltaic cells.

Excitation voltages:
- +5, +12, or +15Vdc ($\pm5\%$). For maximum current requirements, see Table 1. Units with internal cycle interpolation require a maximum of 275ma.

Output format:
- Two count channel outputs (A and B) in phase quadrature with an optional zero reference (ZR) output.

Quadrature specification:
- Units without cycle interpolation: $90° \pm 20°$ (at 10KHz output frequency).
- Units with cycle interpolation: $90° \pm 45°$ (at 10KHz output frequency).

Symmetry specification:
- $180° \pm 10°$ (at 10KHz output frequency).

Rise and fall times:
- 1 μsec (maximum) into 1,000pf load capacitance (applies to squarewave output units only).

Frequency response:
- Units without cycle interpolation: 100KHz for count channels, 50KHz for zero reference (stated at count channel frequency).
- Units with cycle interpolation: Varies according to cycle interpolation factor as follows: 2X — 180KHz/90KHz, 4X — 360KHz/180KHz, 5X — 450KHz/225KHz.

Zero reference angular width:
- $1 \pm \frac{1}{4}$ count channel cycle. Units with line driver output or internal cycle interpolation provide a $\frac{1}{4}$ cycle wide gated zero reference designated as GZ.

Zero reference alignment:
- Center of ZR signal aligns between 70° and 190° of Channel A output at 10KHz output frequency.

Phase sense:
- Channel A leads Channel B for counterclockwise rotation of the shaft as viewed from the shaft end of the unit.

Pin connections:
- For units with connectors, see Tables 2 and 3. For units with cable egress, see Table 4.

Output specifications:
Waveform:

Sinewave
(See Figure 1)
- **Signal levels:**
- Count channels: Sinewave outputs with amplitudes of 30mv p-p (minimum) into a 2KΩ load at 40KHz output frequency. DC offset is $\pm10\%$ of p-p (maximum) signal output.
- Zero reference: 20mv (minimum) usable signal level into a 2KΩ load at 40KHz count channel output frequency.

Squarewave
(Units without internal cycle interpolation - see Figure 2)
- Count channels: 2.5 volts ($\pm.5$ volts) p-p differential sinewaves. Zero reference: TTL pulse with center aligned between $90° + 180°$ of channel A and width of $360° \pm 45°$.
- 5 volt units without line drivers: TTL compatible outputs from a 7404* output stage with 16ma sink current.
- 5 volt units with line drivers: Output stage is an 8830* differential line driver with 40ma sink and -40ma source current.
- 12 and 15 volt units without line drivers: Output levels are determined by the value of user supplied pull-up and load resistance. Output stage is an open collector 7406* with 40ma/30V capability.
- 12 and 15 volt units with line drivers: Output stage is an 88C30* differential line driver with 22ma sink and -40ma source current.

Squarewave
(Units with internal cycle interpolation - see Figure 2)
- Available in 5 volt units only. TTL compatible complementary outputs from an 8830* output stage with 40ma sink and -40ma source current.

Output options:
- Reversed phase sense — Channel B leads Channel A for counterclockwise rotation as viewed from the shaft end of the unit.
- Custom electronics can be provided for a non-recurring charge.

MECHANICAL

Outline dimensions:
- See Figures 3, 4, 5, and 6.

Shaft loading:
- 40 lbs. axially, 35 lbs. radially (maximum).

Shaft radial runout:
- .001" T.I.R.

Starting torque at 25°C:
- Models without shaft seal: 2.0 oz.-in. (maximum).
- Models with shaft seal: 5.0 oz.-in. (maximum).

Shaft angular acceleration:
- 10^5 radians/sec² (maximum).

Moment of inertia:
- 4.5×10^{-4} oz.-in.-sec².

Bearing type:
- Sealed ABEC class 5 or better (varies with resolution capability)

Connectors:
- MS3102R18-1P (10 pin) connectors can be supplied with any output configuration. MS3102R16S-1P (7 pin) connectors can be supplied only with the following output configurations (channels): sinewave (A^+, A^-, B^+, B^-), single ended squarewave (A, B, ZR), complementary squarewave (A, $\overline{A}$, B, $\overline{B}$) and line driver (A, $\overline{A}$, B, $\overline{B}$). MS3106A18-1S (10 pin) and MS3106A16-1S (7 pin) mating connectors are available at additional cost.

Cable description:
- Individually shielded twisted pairs plus an overall shield. Cable contains 10 conductors. Unused wires will be clipped back to overall shield.

Bearing life:
- 2×10^8 revolutions (minimum) at rated shaft loading. 5×10^{10} revolutions at 10% of rated shaft loading (manufacturers specification).

Housing material:
- SR319 cast aluminum.

Cover material:
- Terne coated steel.

Shaft material:
- 303 series stainless steel.

Maximum operating speed:
- 3,000 RPM or maximum count channel output frequency, whichever occurs first.

Slew speed:
- 5,000 RPM (maximum).

Shaft seal [optional]:
- Garlock Type 63. (Operation above 3,000 RPM with shaft seals is not recommended.)

Weight:
- 17 oz. (maximum).

Error:
- See pg. 6.

ENVIRONMENTAL

Operating temperature range:
- 0° to +70°C.

Storage temperature range:
- -25° to +90°C.

Shock:
- 50G's for 11 milliseconds duration.

Vibration:
- 20Hz to 2,000Hz at 20G's.

Humidity:
- To 98% R.H. (non-condensing).

or performance equivalent.
SPECIFICATIONS SUBJECT TO CHANGE WITHOUT NOTICE.

56

FIGURE 1

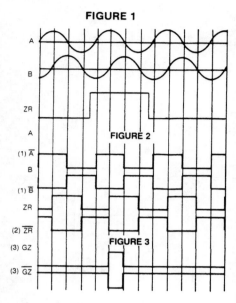

FIGURE 2

FIGURE 3

- NOTES -
(1) Provided only with line driver , cycle interpolation, or complementary output units
(2) Provided only with complementary output units
(3) Provided only with line driver or cycle interpolation output units

TABLE 1					
MAXIMUM CURRENT REQUIREMENTS (milliamperes)					
Supply Voltage	Sinewave Output Units		Output Circuitry 3, 7		Line Driver Output Units
	NoZR	ZR	NoZR	ZR	NoGZ	GZ
+ 5v	85	170	125	220	175	245
+ 12v	85	170	170	170	110	110
+ 15v	85	85	170	170	110	110

TABLE 2				
MS3102R16S-1P (7 PIN) CONNECTOR PIN DESIGNATIONS				
Pin	Sinewave A⁻, A⁻, B⁻, B⁻	Single Ended Squarewave A, B and ZR	Complementary Squarewave A, $\overline{A}$, B, $\overline{B}$	Line Driver A, $\overline{A}$, B, $\overline{B}$
A	Channel A+	Channel A	Channel A	Channel A
B	Channel B+	Channel B	Channel B	Channel B
C	Channel A-	Channel ZR	Channel $\overline{A}$	Channel $\overline{A}$
D	+ Vdc	+ Vdc	+ Vdc	+ Vdc
E	Channel B-	Spare	Channel $\overline{B}$	Channel $\overline{B}$
F	Common	Common	Common	Common
G	Case Ground	Case Ground	Case Ground	Case Ground

TABLE 3			
MS3102R18-1P (10 PIN) CONNECTOR PIN DESIGNATIONS			
Pin	Sinewave Outputs	Squarewave or Line Driver Outputs	Interpolation Units
A	Channel A+	Channel A	Channel A
B	Channel B+	Channel B	Channel B
C	Channel ZR+	Channel ZR or GZ	Channel GZ
D	+ Vdc (±5%)	+ Vdc (±5%)	+5 Vdc (±5%) only
E	Spare	Spare	Test Waveform
F	Common	Common	Common
G	Case Ground	Case Ground	Case Ground
H	Channel A-	Channel $\overline{A}$	Channel $\overline{A}$
I	Channel B-	Channel $\overline{B}$	Channel $\overline{B}$
J	Channel ZR-	Channel $\overline{ZR}$ or $\overline{GZ}$	Channel $\overline{GZ}$

HOW TO ORDER

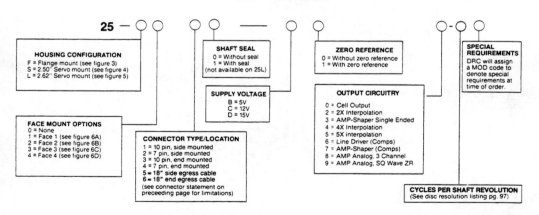

25 —

HOUSING CONFIGURATION
F = Flange mount (see figure 3)
S = 2.50" Servo mount (see figure 4)
L = 2.62" Servo mount (see figure 5)

SHAFT SEAL
0 = Without seal
1 = With seal
(not available on 25L)

ZERO REFERENCE
0 = Without zero reference
1 = With zero reference

SPECIAL REQUIREMENTS
DRC will assign a MOD code to denote special requirements at time of order.

SUPPLY VOLTAGE
B = 5V
C = 12V
D = 15V

OUTPUT CIRCUITRY
0 = Cell Output
2 = 2X Interpolation
3 = AMP-Shaper Single Ended
4 = 4X Interpolation
5 = 5X interpolation
6 = Line Driver (Comps)
7 = AMP-Shaper (Comps)
8 = AMP Analog, 3 Channel
9 = AMP Analog, SQ Wave ZR

FACE MOUNT OPTIONS
0 = None
1 = Face 1 (see figure 6A)
2 = Face 2 (see figure 6B)
3 = Face 3 (see figure 6C)
4 = Face 4 (see figure 6D)

CONNECTOR TYPE/LOCATION
1 = 10 pin, side mounted
2 = 7 pin, side mounted
3 = 10 pin, end mounted
4 = 7 pin, end mounted
5 = 18" side egress cable
6 = 18" end egress cable
(see connector statement on preceeding page for limitations)

CYCLES PER SHAFT REVOLUTION
(See disc resolution listing pg. 97)

*See figure 5 for cable egress locations and dimensions. Cable egress can be provided with any housing configuration.

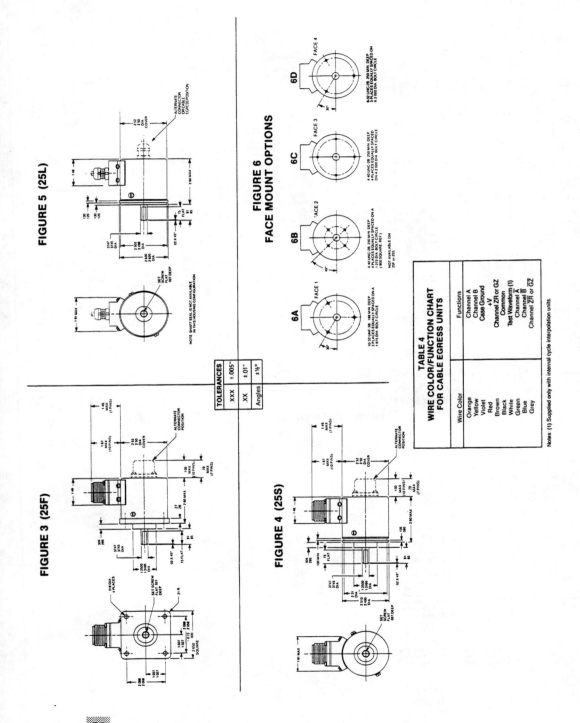

FIGURE 5 (25L)

FIGURE 3 (25F)

FIGURE 4 (25S)

FIGURE 6
FACE MOUNT OPTIONS

TOLERANCES	
.XXX	±.005"
.XX	±.01"
Angles	±½°

6A FACE 1

6B FACE 2

6C FACE 3

6D FACE 4

TABLE 4
WIRE COLOR/FUNCTION CHART
FOR CABLE EGRESS UNITS

Wire Color	Functions
Orange	Channel A
Yellow	Channel B
Violet	Case Ground
Red	+V
Brown	Channel $\overline{ZR}$ or $\overline{GZ}$
Black	Common
White	Test Waveform (1)
Green	Channel $\overline{A}$
Blue	Channel $\overline{B}$
Grey	Channel $\overline{ZR}$ or $\overline{GZ}$

Notes: (1) Supplied only with internal cycle interpolation units.

58

There must be a detailed set of specifications. I was contracting with National Semiconductor Corp development of a custom chip. We worked together on cations, which ended up as an eighteen page definition of what was to be provided as a device.

We should have a detailed set of product or device specifications, even when there is no custom chip in the design. Otherwise, we really do not have an understanding between the principals of exactly what it is that is to be done.

5. Establish two categories: the MUSTS and the WANTS.

These categories will

Help us to prioritize.

Help us to clarify the objectives.

Help us to decide whether there should be two products instead of one; perhaps there should be a relatively simple product and a relatively complex product in the product line.

Help us decide how to limit the "bells and whistles"

6. We are making sure that *everyone* on the team has clearly in mind and has written down on paper (documented) just exactly what it is that is to be accomplished.

7. Hold a two-hour meeting every day or at least every other day until the product definition is detailed, accurate, and complete.

Some individual should record all the important points, and, at the end of each meeting, provide copies, in the rough, for everyone at the meeting.

The next meeting should be scheduled before adjourning.

8. Write the software first, in a software driven product. The software characterizes the end product for the user more so than the hardware, and it is important to focus on end user requirements.

9. Write the Operating Manual first. The Operating Manual describes the user interface, and, again, it is important to focus on end user requirements.

The goal is to make *sure* the definition is detailed, accurate, and complete.

It is extremely important to *gain concurrence* by everyone involved with the product that the product definition is correct and complete.

10. Loop top-down, bottom-up, and top-down as illustrated in Figures 2.1 and 4.4.

11. When complete, everyone on the team signs and dates the detailed product specification.

Do not sign the functional specification until after the time and cost schedules have been established as discussed in Chapters 3–5.

THE PROJECT DESCRIPTION

The first two elements of the product plan, the functional specification and the detailed product specification, are elements that describe the product. The company-wide, cross-functional team fully described the product to be developed. Further, in the process of doing so, the team gained everyone's concurrence. Now it is time to describe the *project* of developing the new product. The first two elements describe the product; this third element describes the project. The project description provides, for all members of the team, the definition of the development project.

Finalizing the specifications is a prerequisite for writing the project description. This description, like the functional specification, is relatively brief. Actually, the fourth and fifth elements of the product plan, the plan network and the resource allocation, also describe the project and, when they are detailed, in later sections of this book, the project description will be more complete. We must first describe the development project.

The basic project description may only be one page long. An example is given in Figure 2.4. It is a document that summarizes the important points derived in the process of finalizing the specifications. Furthermore, it lists the team members by name, and it specifies who the project leader will be, by name. *It is a milestone report to higher management.* It states the time and cost estimates and states, furthermore, that these estimates will be improved by the process of generating the plan network, the resource allocation, and the cost schedule, which are the remaining elements of the overall plan. The project description, additionally, indicates to higher management when these other documents will be completed for review. The project description states that the project will be carried out in accordance with the approved functional specifications and detailed product specifications.

PROJECT DESCRIPTION
for the
HOME

This project is the construction and development of a home. It includes the house, appurtenant structures, and all architectural landscaping. All preparations for occupancy will be completed.

The project will be carried out in accordance with the functional specifications and approved detailed product definition, both of which are attached. Final checks will be measured against the design specifications by the number.

Current estimates are that the total cost of the project will be $125,000 and that the completion date will be (date). These estimates will be improved in the process of developing the plan network, resource allocation, and cost schedule. The network, allocation, and cost schedule will be developed, and a report submitted on or before (date)

The project team is as follows:

	Name	Date
Project Leader:	_____	_____
	_____	_____
	_____	_____

FIGURE 2.4 Project description for the home.

The *Overall Procedure* states that 40–50 activities should be listed in the process of writing the project description. These need not be listed in the project description document. They are needed, though, for the thought process leading to the project description.

THE PLAN NETWORK

The plan network is provided for use as a planning, scheduling, and controlling management tool. It is an overall graphical illustration of the project. We will use it much as travelers in an automobile on a cross-country journey use a road map. A significant number of people from industry have told me that their companies do not use a plan network—Gantt charts, or any other good graphical illustration of their projects. These people from industry are eager to learn of the plan network and its use as a project management tool.

Can we imagine travelers on an automobile cross-country journey without road maps? No, if we are the travelers, we would make it a point to obtain good maps, up to date, complete and detailed. Similarly, as we set out on a product development, we want an accurate, complete, detailed, graphical illustration of the project.

We should think of the plan network as the project. True, it is a chart consisting of arrows and circles, but more than that, it is the project, all on one piece of paper or two. With this interpretation it will be very useful to us.

There are several versions of project networks now in use, and many of them are computerized. PERT charts and Critical Path Method charts are used. Gantt charts are preferred by some, and the Mechanized Gantt chart is promoted in some circles. MacProject, Super Project, Timeline, Primavera, and many others are available. We will study, herein, the fundamental concepts, requirements, and characteristics of network planning. The reader can apply these fundamentals to any manual or automatic process.

The plan network is an overall graphical illustration of the development project.

I have found, in working with over 150 people from industry over the last four years, that not having a project planning software package for network planning is not the fundamental problem. Rather, it is that people in industry do not have an understanding of the fundamentals of network planning. In many cases, their companies forge ahead with their development projects without *any* detailed plan on paper. These people, with whom I have worked, are eager to gain an understanding of how to list project activities and prioritize and graph them in a meaningful order. Learning about these fundamentals and how to adapt them to their projects is their real interest. Once they have mastered the fundamentals and the background motivation, then the acquisition and use of a software package is easy.

We will, therefore, spend time and space here discussing the fundamental concepts, requirements, and characteristics of network planning. The reader will then be qualified to use any computerized program of choice.

The presentation here will be a modernized version of PERT, which is preferred to Gantt because it shows more clearly the interrelationships between the serial and parallel activities. Secondly, most PERTs in other presentations are drawn so that there are more or less uniform distributions of the activities and activity completion events spread evenly across the chart paper. That is, most PERTs are not drawn on a linear time scale. This author prefers to not only present a clear chart but also to draw the chart against a linear time scale. Then, when the cross-functional team refers to the PERT and uses it as a project control tool, the team can easily match actual progress to expected progress in terms of calendar dates. The traditional Critical Path Method (CPM) chart is an inverted version of the traditional PERT chart, and is not preferred by the author. It is a simple task for anyone to translate a PERT chart to a CPM chart, if desired. However, the Critical Path Method as a special sequence of crucial activities is important. It will be identified and analyzed as part of our PERT chart study; the Critical Path Method is a very important subset of the overall network of activities graphically illustrating the complete project.

We will draw our graphical plan on a linear time scale to make it easy for us to align with the calendar and with our resource allocation and cost curve.

Later, after the plan network, resource allocation, and cost schedule are generated, all three can and should be drawn against the same time scale.

The plan network must be prepared with great care. It must represent the product development project in all the details spelled out in the functional specifications and in the detailed product specifications. It must be in close conformance to the available resources. It must illustrate that as many activities will be carried out in parallel as possible and, yet, it must be realistic. The team is under obligation to carry out the development project as rapidly as possible to reduce the time to market. However, the team is expected not to make promises it cannot keep in terms of completion dates—with the limitations in resources.

The test is: Can anyone, the supervisor or higher-level manager, another team, group, or individual do better? Is it possible to present a plan network that illustrates a faster product development that is equally limited by resources—by doing more activities in parallel? If the answer is no, then the cross-functional team has prepared the best network possible, and it is time to move forward with preparing the work breakdown structure and the cost.

The team should spend plenty of time on the plan network and, yet, stop at the point of diminishing returns. It is important to move forward at the right point in the evolution of the plan network. We should not spend undue time drawing plan networks. Yet, it is extremely important that we generate one and make sure it is the best one possible at this point in the evolution of the overall plan. Suffice it to say that studies have proven that 70 percent of the reported project delays have been the result of inadequate planning at the outset. [1]

It is extremely important to have

1. agreed upon functional specifications,
2. agreed upon detailed product specifications,
3. the correct product design concept, and
4. the best possible plan network.

We remember, at this point, that the plan network is a preliminary draft. As we structure the work breakdown in Chapter 4, we will be subject to resource limitations, and the plan network will then be refined accordingly.

We must remember that the correct product design concept is all-important. The design concept determines the relative usefulness to the end user. The concept of the product design determines the cost of product, the producibility, the product reliability, the amount of capital equipment required to produce it, the amount of floor space required in production, the ease with which the product can be serviced, and the cost of product development. Therefore, we must first make sure that we have the optimum product design concept. Then the plan for developing the product, including the plan network, is based on this optimum design concept. So, it is important for the cross-functional team to be highly creative in conceptualizing the design. We remember the creativity of the physics student in Chapter 1.

The correct product design concept is all-important.

The plan network should consist of approximately forty activities. It is possible to draw a one-activity network: Develop the Product. It is also

---

[1] Ashok K. Gupta and David L. Wilemon, "Accelerating the Development of Technology-based New Products," *IEEE Engineering Management Review* (December 1990).

possible to network illustrate 1,000 activities if the team were to include the procuring and fastening of all the hex nuts! It was stated earlier that the network was to be used as a planning, scheduling, and controlling management tool. The team must stop, in the preparation of the network, at the point of diminishing returns. We want the plan network to save us time in the end, not to waste time. It is recommended that at least thirty activities be included along the critical path. Then, the implication is that the project can be controlled to an accuracy of 2 or 3 percent.

The in-depth basics of network planning will be given here. Consequently, the reader will understand all the elements of network planning, scheduling, and controlling, and can exercise these management components either manually or automatically. The presentation will be program invariant; the teachings herein can be applied to any automatic, computerized program, such as those identified previously.

Many times people suggest that feedback paths and loops be drawn as a part of the plan network to indicate the dynamics of re-doing certain activities until all the requirements are met. For example, it is understood that, in carrying out research, often more than one iteration must be taken before adequate results are achieved. We understand that more than one iteration must be taken in many cases. However, in the interest of keeping the plan network chart neat, all iterations are depicted in one forward path. We will not illustrate the backward looping activities that actually take place; it will be understood that all these dynamics are included in the forward paths.

The scheduled time for the activities will not be entered on the chart, at this point. As pointed out earlier, this must wait until Chapter 3.

The elements of the plan network will be explained in detail as the examples are presented. The explanation will be detailed enough to prepare the reader to not only completely understand the examples given but to also be fully trained to generate a plan network for any project he or she will undertake next.

Several basic network structural elements and rules will now be given. These are standard elements and standard rules as applied to PERT charts as distinguished from CPM charts, Gantt charts, and other illustrations representing projects.

The PERT chart is a set of lines and circles all connected together in a unique way for a given project. The project is broken down into forty or fifty activities—or more if the project team so chooses. Each activity, except for the first one, has a requirement for other activities to have been completed prior to its commencement. That is, the team will decide that there is a sequence in which the activities should be carried out. As each activity is considered, it will be understood that some activities must be completed before the one under consideration can begin. So there is a string of sequential activities.

Figure 2.5 is an illustration of the way in which the project activities and completion events are presented in graphical form. The symbols used are lines and circles. Each activity is represented by a line, and these are connected on the network diagram by circles. The line represents activity duration time and each circle represents a completion event. Each event is a point in time. For example, in Figure 2.5, the circle between Activities b and c represents the point in time at which Activity b is completed. It also signifies that Activity c can begin at that point in time, not before.

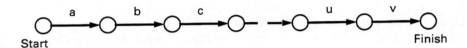

Activity b can begin only after Activity a has been completed.

FIGURE 2.5 The concept of network planning illustrated.

Then it will be discovered that many activities can be carried out in parallel. *Care must be taken here to distinguish between activities that must be carried out sequentially because of the structure of the product, whatever it is that is being developed, and those that must be carried out sequentially because of limited resources.* Many project teams or project leaders fail to schedule activities sequentially where it is clear, with adequate study, that some activities which could otherwise be done in parallel simply cannot be done that way because there are not enough talented people available at the same time.

Example 1

BUYING AN AUTOMOBILE

For example, the simple project of buying an automobile consists of the following activities:

a decide to buy used car

b set dollar limit

c visit ten dealers and lots

d narrow search to three top candidates

e decide to buy car of choice

f acquire funds

g mechanic's check

h sign for car and take delivery

A recommended procedure is to draw a rough sketch plan network based on activity sequencing caused by product structure only, with disregard to resource limitations. This technique enables the team to get on with it—to get a complete configuration down on paper. Then the required resources are listed *and checked out.* As a result of checking resources, it is most often found that it will be impossible to run as many activities in parallel as first thought. Consequently, some of the "parallel" activities must be done in sequence because of these resource limitations.

For example, in Figure 2.6, we first list all the activities in series.

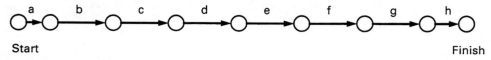

FIGURE 2.6 All activities in series.

Then, we will expedite the process by planning on having two people search for the automobile, and, also, we can arrange for the funding in parallel with some of the other activities. We illustrate this in Figure 2.7.

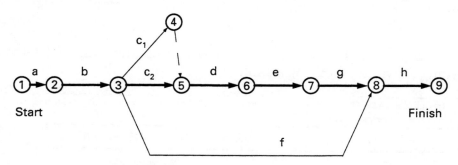

FIGURE 2.7 Some activities in parallel.

We notice a dashed line in Figure 2.7; it is merely an indication that Activity c_1 must be complete, as well as Activity c_2, before d can begin. That is, we must search through the dealers' lots, according to our plan, before we can narrow the list of possible cars to three candidates. You ask, "Why not run the line representing Activity c_1 directly from Event 3 to

Event 5?" The reason is that software programs, used for automating this process, will then have listed two activities, each of which runs from Event 3 to 5, and the computation will stop. Programming is done by identifying the activities, in each case, by the preceding and succeeding event numbers. That is, Activity a would be input as Activity 1–2 and Activity b is input as Activity 2–3. So, having two activities with identical preceding and succeeding event numbers is too confusing to the computer and, therefore, is not allowed.

The solution is to use a so-called Dummy Activity. Then, we understand that it is merely a connector indicating that we must complete Activity 3–4, as well as Activity 3–5 before we can begin Activity 5–6. Dummy activities are also used to connect multiple projects, as explained in Chapter 7.

We return, now, to the process of generating the plan network. We find, upon asking people to search for a used automobile, that only one person is available. Therefore, we must modify our plan accordingly. Figure 2.8 illustrates the final plan.

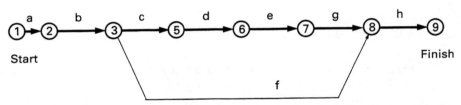

FIGURE 2.8 Final plan network for project of buying an automobile/parallel activities limited by resources.

Table 2.1 lists the activities and the corresponding, immediately preceding activities for buying an automobile.

TABLE 2.1 Activities and corresponding immediately preceding activities for buying an automobile.	
ACTIVITY	IMMEDIATELY PRECEDING ACTIVITY
a	None
b	a
c	b
d	c
e	d
f	b
g	e
h	f,g

As another example, suppose a product development requires two printed circuit boards to be developed. If there are enough circuit designers, board layout people, and test technicians, then the two boards can be developed in parallel. However, if there is only one designer and/or one workstation layout person, or one workstation and/or test technician, at least some of the activities are going to have to be carried out sequentially. Another example is that if a highway is being developed and there are two different bridges that must be designed into the highway, then, whether the two bridges can be designed simultaneously depends heavily on whether there are two bridge design teams available.

The procedure we will follow is:

1. The cross-functional team will meet and list all forty or so of the activities composing the project of developing the new product.
2. The activities are arranged roughly in the order in which they would be carried out. It is not necessary that this sequencing be highly accurate at this point.
3. List adjacent to each activity the other activity or activities that must immediately precede the activity under consideration.
4. Translate this list into a rough plan network chart. Disregard resource limitations, for now. Remember that this is a first approximation and it enables us to get on with it. *We must remember, however, that resource limitations must be considered before the plan network is finalized.*

A unique series-parallel chart configuration derives from the list of activities, corresponding preceding activities, and resource considerations. The chart consists of lines and interconnecting circles. Each line represents an activity and a corresponding time period—the period of time over which the activity is carried out. Each circle represents a start or a completion event—a point in time.

Example 2

THE DEVELOPMENT OF A PRODUCT, THE GENERAL CASE

To further illustrate the process of considering a project, breaking it down into a number of activities, and prioritizing them, we will use the following example. We will generate a network configuration that will serve as a graphical illustration of all the series and parallel activities composing the project. Let us choose, as a project, the development of a product, in the general sense.

Table 2.2 is the list of some of the activities we must carry out in the development of a product. Given a specific product, our *Overall Procedure* indicates that we would, by now, have in our hands the detailed product definition, and we would be dealing with the specific definition of the product under development. However, our purpose here is to familiarize ourselves with network planning, and we treat the problem in the more general sense.

The two lists, activities and corresponding immediately preceding activities, enable us to configure the project.

The team, once having determined the activity priority as tabulated previously, can now sketch out the order in which the activities will be carried out—in graphical form—illustrating the time relationships between all the activities. The result is given in Figure 2.9.

TABLE 2.2 Activities and preceding activities for product development.

ACTIVITY		IMMEDIATELY PRECEDING ACTIVITY
DS	Design system	Start event
PD	Preliminary design of product	DS
CP	Construct prototype for feasibility testing	PD
TF	Test feasibility	CP
OM	Write operator's manual	DS
DDP	Design and develop the product	TF,OM
DMP	Design and develop the manufacturing processes	TF,OM
DMF	Design and develop the manufacturing facilities	DMP
DQE	Design and develop the quality engineering procedures	TF,OM
TE	Procure test equipment	DQE
PC	Procure components for company-wide prototypes	DDP,DMP,DQE
CP	Construct prototypes	PC,DMF
TP	Test prototypes	CP,TE
C	Estimate cost	CP,TE
BT	Carry out Beta testing	TP,C
	Go into production	BT

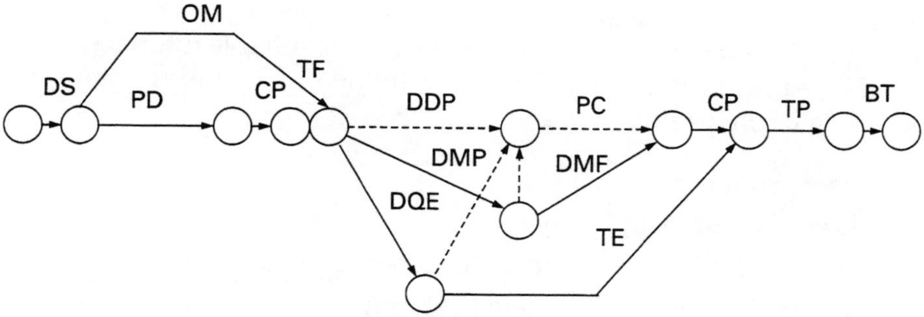

FIGURE 2.9 Plan network for development of a product.

Example 3

DESIGNING A MAJOR COMPONENT PRODUCT INTO A SYSTEM

So far we have given a simple, eight-activity example and a generalized product development example. We follow these, now, with an example that further illustrates project management. It is the very typical case of designing a major component into a system. This may be a new system design, where the major system component is purchased. It may be a replacement project, necessitated by the component's predecessor having gone obsolete. Or, the major component may be provided by a project within the company. The company may decide to design and develop the product to be integrated into a larger system or into a product line. Whatever the case may be, the project considered here is the same: design the component product into its larger system. The correct product must be selected, input signals may have to be formatted, interface requirements must be met, and, if it is a software-driven product, its computer will have to be programmed. Finally, we must test and document the resulting system.

In this example, it has come to our attention in the company, let us say, that the printer in a hard-copy output system must be replaced. This example covers the case of product development in a manufacturing company. It may be new product development. Or, it may be the case of product upgrading or maintenance. We will assume, in the illustration, that a printer in a printer system will no longer be available from the supplier. Therefore, a different printer must be made available for the company to continue to supply the printer system. The reader may reinterpret the problem if, in the reader's situation, a printer is to be designed and developed within the company. In the illustration, we will assume that another one can be found in the marketplace, and the project will be to integrate it into the overall system by designing suitable interfaces.

The team has determined that the following activities compose the upgrading of the printer system:

Select printer
Assemble and install printer
Generate databank for output form boiler plate
Write main printer processing program
Design Input A signal format
Design Input B signal format
Design Input A interfacing board
Design Input B interfacing board
Finalize documentation for Channel A
Finalize documentation for Channel B
Test system
Document system and system test

To determine the order in which the activities must be carried out, it is necessary to study priorities and constraints. Through study, we establish the following facts:

1. Writing the printer processing program depends only on having selected the printer.
2. Selecting the printer must, of course, precede printer installation.
3. No program can be written, nor interface board designed, until the particular printer model has been selected, although the input signals can be formatted while alternate printers are evaluated.
4. The interface boards cannot be designed until the corresponding signals have been formatted.
5. The printer processing program and Interface Board A are required to establish the databank, which can be created on rental equipment.
6. Documentation for each channel can begin after signal formatting.
7. The final activity is system testing.

From the above facts we can structure the order in which each and every activity must be carried out. This process results in the tabulation of corresponding immediately preceding activities, shown in Table 2.3.

TABLE 2.3 Printer replacement activities and precedence.

	ACTIVITY	EXPECTED TIME	IMMEDIATELY PRECEDING ACTIVITY
a	Select printer	2 weeks	None
b	Assemble and install printer	6	a
c	Design Input A signal format	2	None
d	Design Input B signal format	1	None
e	Write main printer processing program	6	a
f	Design Input A interface board	5	a,c
g	Design Input B interface board	4	a,d
h	Generate databank for output form	3	e,f
i	Finalize documentation for Channel A	4	e,f
j	Finalize documentation for Channel B	4	a,d
k	Test system	2	b,g,h
l	Document system and system test	1	b,g,h

Having established the activity precedence, we can now draw the plan network configuration and number the events. The tabulation of activities and corresponding immediately preceding activities completely specifies the order in which the project is to be carried out. From this tabulation we can draw the network configuration, which graphically illustrates the series and parallel activity interrelationship. Again, we follow the rules presented at the beginning of The Plan Network on page 61.

The plan network for the printer replacement project is given in Figure 2.10.

We will compute the activity duration times and completely schedule the printer project in Chapters 3 and 4. Resources will be allocated in Chapter 4.

The printer replacement project can be interpreted to be a full product development project. All the elements of product scheduling are embodied in this illustrative example.

We must remember that in a real project we want an activity breakdown that results in tabulating/graphing forty to fifty activities. We limit our example here to a smaller number of activities for simplicity of illustration.

The next example, the development of a home, is an illustration of a project made up of forty-five activities.

73

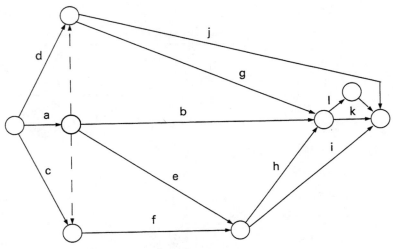

FIGURE 2.10 Printer replacement plan network.

Example 4

DEVELOPMENT OF A HOME, A CONSTRUCTION EXAMPLE

The first example we choose to detail is that of developing a home. We wish to detail the Overall Procedure of developing a product including a full treatment of each of the seven major elements in the overall procedure. The home is a product that begins with land that is then developed and improved. Many product developments in industry follow this same process. The development of a home involves selection of location and land, landscape design, site identification, construction of the house, construction of any appurtenant structures, driveway and sidewalk installation, planting of trees and shrubs, and landscape development. So, the development of a home is much more than the construction of a house. This is being pointed out here to emphasize that most product developments in industry, whether the construction industry or elsewhere, are not simply the building of a product. Many developments are more complex. They include the design and building of the product, development of operating and instruction manuals, development of ancillary products such as special tools and fixtures to be used in conjunction with the main product, development of software that the operation of the hardware product depends on, and so forth.

We will now construct the plan network for a full, forty-five–activity project. We will do it in sections to facilitate understanding. The project will be that of procuring land, building a house, and developing the complete home, including the landscaping and other land development.

The reasons the author has chosen the development of a home as the full project example in this book are the following:

1. Everyone is quite familiar with the home; the reader can concentrate on the elements of project management without getting hung up on the concept and structure of the product.

2. Building a home is a valid example of a technical project, especially with the modern heating, ventilation, and air conditioning systems; security systems; and advanced materials used for insulation and other features.

3. Developing a home is a project of moderate complexity. Enough parallel and serial activities are involved that it provides good training to the student on planning, scheduling, and controlling.

The first fifteen to twenty activities that randomly come to mind are:

Frame walls	Install roof trusses
Roof sheathing	Fireplace and chimney
Install temporary power	Interior finishing
Pour foundation and basement walls	Identify land for home
Excavate	Install HVAC system
Exterior finishing	Plumbing
Shingle roof	Wiring
Survey and locate foundation	Landscape design

This first list should include all the high-priority main categories of the activities composing the project—not the details. For example, excavation; foundation; framing; plumbing; heating, ventilation, and air conditioning; and the electrical system should be included in this first pass. Installing light bulbs, testing the garage door opener, and planting flowers would not be included in this first listing of the main categories.

In the second-level listing we would include excavation for the house, excavation for the garage, and excavation for the shed, foundation for the house, foundation for the garage, foundation for the shed, framing of the house, framing of the appurtenances, electrical wiring for the 110 volt circuitry, for the 220 volt circuitry, and for 24 volt circuitry, and so forth.

We would continue this process until we have thirty or so activities on the critical path and forty or so all together.

Reordering the first list of main categories and listing them along with the respective preceding activities results in Table 2.4.

TABLE 2.4 The list of activities and preceding activities.

ACTIVITY NO.	ACTIVITY NAME	PRECEDING ACTIVITY
a	Identify land for home	—
b	Landscape design	a
c	Survey and locate foundation	b
d	Excavate	c
e	Pour foundation and basement walls	d
f	Frame walls	e
g	Install roof trusses	f
h	Plumbing	e
i	Roof sheathing	h,g
j	Fireplace and chimney	e
k	Shingle roof	i
l	HVAC	k,j
m	Wiring	l
n	Interior finish	m
o	Exterior finish	n

We now draw a rough network of the fifteen or twenty activities. This is to get organized. It is a diagram of the flow of activities. Figure 2.11 is a graphical illustration of the project insofar as these few activities describe the project. Later on, we will present a larger network with more than forty activities.

In drawing the plan network, we use the Activity/ Precedence Table 2.4. Activity a obviously is the starting activity, since nothing precedes it. Then Activity a precedes Activity b, and so on.

Activity i is preceded by both Activities h and g. Therefore, the network is drawn to indicate that both these activities must be completed before Activity i can begin.

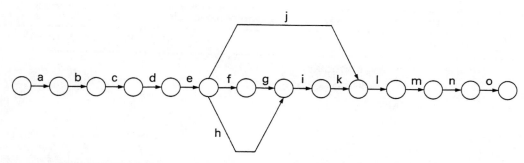

FIGURE 2.11 Plan network of major activities for development of home.

So far, the network does not take into account possible resource restrictions. Nor has there been any consideration for the amount of time each activity takes. These matters will be settled later. Suffice it to say, at this point, that there is now a relatively clear graphical illustration of how the new product development will be structured—so far as the fifteen to twenty top-level activities are concerned.

Next, we list *all* the activities composing the development project and the corresponding preceding activities. This list is presented as Table 2.5.

TABLE 2.5 List of all activities with corresponding preceding activities.

ACTIVITY		PRECEDING ACTIVITY
1–2	Identify land for home	—
2–3	Landscape design	1–2
3–4	Survey and locate foundation	2–3
4–5	Excavate	3–4
5–6	Underground drainage	4–5
5–9	Install temporary power	4–5
6–7	Pour foundation	5–6
7–8	Insulate basement	6–7
8–9	Backfill	7–8
9–10	Plumbing service	5–9,8–9
10–11	Basement plumbing	9–10
9–13	Frame walls	5–9,8–9,
13–14	Install wall sheathing	9–13
14–15	Install roof trusses	13–14
11–12	Inside plumbing	10–11

TABLE 2.5 List of all activities with corresponding preceding activities (cont.).

ACTIVITY		PRECEDING ACTIVITY
12–15	Plumbing inspection	11–12
15–16	Roof sheathing	14–15,12–15
16–17	Shingle roof	15–16
17–18	Install solar system	16–17
9–19	Fireplace	5–9,8–9
19–20	Chimney and chimney flashing	9–19,17–18
18–20	Pour basement floor	17–18
20–21	Install HVAC system	18–20,19–20
20–24	Windows and exterior doors	18–20,19–20
24–28	Install siding	20–24
21–22	Wiring	20–21,20–24
22–23	Security system	21–22
23–29	Wiring Inspection	22–23
20–25	Grade lot	18–20,19–20
25–26	Driveway and sidewalk	20–25
26–27	Plant shrubs	25–26
27–28	Lay sod	26–27
29–30	Insulate walls and ceilings	23–29,24–28,27–28
30–31	Sheet rock	29–30
31–32	Taping and skim coat	30–31
32–34	Install cabinets	31–32
32–33	Install interior doors	31–32
34–35	Lay hardwood floors	32–34,32–33
35–36	Sand floors	34–35
36–37	Finish floors	35–36
37–38	Interior finishing	36–37
38–39	Light fixtures and plumbing fixtures	37–38
28–39	Exterior finish	24–28,27–28
39–40	Clean windows	38–39,28–39
40–41	Clean property	39–40

In Chapter 3 there is a full treatment of scheduling. To the extent that it is extremely important to plan well and prepare a *realistic* schedule, we feel that another whole chapter should be devoted to these subjects.

PROBLEMS

2.1 A prerequisite to good control of a product development project is careful planning at the outset. Imagine the opposite. List the results in some detail. Please limit your answer to one-half page.

2.2 What is the advantage of identifying the cross-functional team *after* the preliminary product specification (see Figure 2.1.) is written?

2.3 Why are the functional specification and detailed product specification written before the project description?

2.4 Why must the plan network be finalized *after* the resource allocation is generated?

2.5 State in fifty words or fewer the difference between the functional specification and the detailed product specification.

2.6 Complete formulation of the functional specification is a top-down, bottom-up process. What is the reason for the second part?

2.7 Why are all cross-functional team members required to sign and date the functional specification?

2.8 Why is it preferable to draw the plan network on a *linear* time scale?

2.9 Draw a five-activity plan network where all activities are in series. All in parallel.

2.10 Beginning with Figure 2.8, draw it on a linear time scale if the activities require the following corresponding times:

$a = 1$ time unit $\qquad$ b and c each $= 5$ units

$d, e,$ and $g = 3$ units $\qquad$ h and f each $= 12$ units

2.11 Modify Figure 2.11 to represent a plan where the roof sheathing is installed prior to the plumbing and the exterior is finished before the interior.

3 CONCURRENT SCHEDULING

INTRODUCTION

Concurrent scheduling is the scheduling of all the activities and tasks composing the overall technical product development project and, in particular, deriving the schedule by working together as a cross-functional team in scheduling these activities in the correct series and parallel order. The cross-functional team is defined in Chapter 1. True, the individual or group of individuals from the function that will eventually do the work in carrying out a particular activity can best provide the schedule for it. However, the cross-functional team must carry out many of the activities, such as the selection of materials of construction of the product under development, the specification of prototype testing, or the time it takes to check out the preproduction model. The whole team must schedule these activities.

They should also be studied by the cross-functional team. Even those estimated by an individual, in each case, and those estimated by small groups of two or three individuals should be studied, afterwards, by the whole team. It is as important for the correct members of the team to give or, at least, to have early, continuous, and effective inputs on manufacturing engineering and quality engineering as it is for the research, design, and development engineering personnel to have inputs to the new product design.

This planning for, scheduling of, and carrying out of the design of the manufacturing processes concurrently with the design of the product by teamwork, and the gaining of the concurrence of *everyone* on the team, is the essence of concurrent engineering. It is important for the design and development of the quality engineering and that of the manufacturing processes to be carried out concurrently with the design and development of the product.

After the activities have all been studied, the plan network should be refined and each activity time listed adjacent to the respective activity.

The objective is to end up with a network diagram of the project that will provide us with a management tool which will facilitate carrying out our responsibilities as managers.

The objective is to end up with a network diagram of the project that will provide us with a management tool. The objective is to have a tool that will facilitate carrying out our responsibilities as managers. Again, the definition of *manager* is this: each and every individual who has a responsibility on the team managing the project. The project leader certainly has a management role. Moreover, we all must manage our time in whatever career we choose, and, in this case, too, we must manage our time to insure maximum return for the individual effort in carrying out our assigned tasks. So, everyone, to that extent, is a manager. This plan network management tool will provide us with an easier way of observing and controlling the status and progress of our project. We must not only observe and evaluate the status and progress completely and accurately; we must *control* the progress of our project completely and correctly. The plan network, realistically scheduled, will facilitate this process.

This plan network management tool will provide us with an easier way of observing and controlling the status and progress of our project.

The procedure so far has been to carry out Steps 1 through 4 in the *Overall Procedure*, which is given on the next page, again, for ready reference. These steps are shown in **boldface** type. We have done the first part of Step 4; more remains to be done.

OVERALL PROCEDURE FOR CONCURRENT PROJECT MANAGEMENT

1. Clarify the *FUNCTIONAL SPECIFICATIONS.* Study the elements of planning to make sure you and your teammates have properly addressed all aspects of planning. Clarify the company objectives and the marketing objectives of the new product development. Study competition. Determine customer's needs. Write product description. Write final functional specifications. Document agreement with the management team by everyone on the team signing the final specifications.

2. Detail the product specifications. Identify and establish the cross-functional team. Study competition. Study the functional specification. Select the design concept. Detail all aspects of the physical product. Rely on experts. Write a *DETAILED PRODUCT DEFINITION.* Document agreement within the cross-functional team by having all members of the team sign the detailed product definition.

3. Describe the project. Elements 1 and 2, above, describe the product. This element, the *PROJECT DESCRIPTION*, describes the project of developing the product. List all the tasks constituting the project—list approximately forty or fifty tasks. Rely on experts. Write a detailed project description. Document agreement within the team in the same manner as above.

4. Generate the *PLAN NETWORK.* List all the tasks/ activities that compose your project. Prioritize them. Indicate after each activity the one or more other activities that immediately precede the one under study. Draw a rough network to illustrate the correct configuration. Rely on experts. Study resources. All members of the cross-functional team must sign the plan network.

5. Validate the **RESOURCE ALLOCATION**. Draw the work breakdown structure and assign the different talents to corresponding tasks. Obtain approval from the source managers for scheduled allocations. Modify the plan network configuration if necessary. Redraw the plan network on a linear time scale. Identify the critical path. All members of the cross-functional team must sign the revised plan network.

   Draw the resource allocation in the form of a histogram/vertical bar chart on the same time scale as used for the plan network. All members of the cross-functional team must sign the resource chart.

6. Generate the **COST SCHEDULE**. Draw a cost curve as a function of time using the same time scale as used for the plan network and the resource histogram. All members of the cross-functional team must sign the cost curve.

7. **REPORT** the plan to management and request funding. Because experts were relied upon for the above planning, the best possible plan has been put in place. The project leader has the responsibility of presenting the plan for approval.

We now develop the network so that it gives the schedule duration for each activity. We will estimate each activity duration after all the aspects of good planning have been properly addressed. Concurrent planning was covered in Chapter 2. This chapter addresses concurrent scheduling.

To begin with, the plan must be based on the correct new product design concept. The design concept is all-important. The degree of success of the product over its life cycle, in terms of performance and profit, depends in large part on the design concept.

We give the following example to emphasize the requirement for conceiving of and pursuing the best design concept. This author was once assigned the project of designing and developing a special purpose computer to blend the various components for lubricating oils and gasolines. An earlier generation computer was based on analog electronic circuitry, and it had limited accuracy and range. I decided to replace all this analog circuitry with digital computer cards properly integrated into an overall special purpose digital computer. This new concept had significantly better accuracy and range, and easily met the functional and detailed product specifications. Furthermore, the change to the digital system led to a successful new product development on a company-wide basis. The analog system was marginal. By using a digital system instead, much time and money were saved throughout the life of the product. The new design concept resulted in a successful product.

Let us turn now to the process of project scheduling.

Activity duration estimates must be made by the people who are going to carry out the respective activities, and these estimates must be edited by experienced people and perhaps modified—with everyone's agreement.

We must base the activity duration estimates on the optimistic, pessimistic, and most likely estimates to obtain more accuracy in planning. If we make just one middle-of-the-road estimate, it will not be as accurate. These estimates must be made by the "workers," and they must be edited by experienced people and perhaps modified—with everyone's agreement. We must gain the concurrence of the workers; we must discuss the schedule with them. If concurrence cannot be gained for a faster schedule, then we fall back to one that everyone thinks is more *realistic*. The schedule must be challenging, that is, we must expedite the completion of the project; however, it cannot be unrealistic.

DEFINITIONS

Some definitions are now in order. We consider three time estimates for each activity in the process of accurately estimating activity duration time.

t_m = *most likely time*

Meaning: If the task were carried out many times by *different* well-qualified individuals and the duration of time required to complete the task by these different people were observed and recorded, the duration of time required to complete the task that occurs most frequently is the most likely time, t_m. (Must be different, skilled people—if the same person were repeatedly tested, then the individual would become exceptionally skilled and the test would be unfair!)

We will estimate t_m first to provide a frame of reference for two other estimates:

t_o = *optimistic time*

Meaning: Probability is at least 95 percent that it will take more time than t_o to complete the task. This means that if a fairly large number of well qualified individuals, say 100, were to each set out to accomplish this task, then 95 out of the 100 would take longer than the time t_o. The remaining 5 would take less time than t_o.

t_p = *pessimistic time*

Meaning: Probability is at least 95 percent that it will take less time than t_p to complete the task. This means that if a fairly large number of well qualified individuals, say 100, were to set out to accomplish this task then 95 out of the 100 would take less time than t_p. The remaining 5 would take more time than t_p.

Then we compute t_e according to the following formula:

$t_e = $ *the expected time* $= (t_o + 4t_m + t_p)/6$

The rationale for using three estimates for each activity is as follows:

By using three estimates, it is possible to take "uncertainty" into account when estimating how long an activity will take.—Gido

A single point estimate assumes certainty. Using three estimates is a better way of handling uncertainty.—Gray

To include uncertainties, a range of variation in job time is provided by the optimistic and pessimistic times.—Phillips, Ravindran, and Solberg

CALCULATING ACTIVITY DURATIONS

Compute the amount of elapsed time it will take to complete each task appearing in the network that has been generated for the specific project. Estimate the elapsed time, not the work time. The time that will be required on the calendar to complete the project will result from the aggregate of the elapsed times. There is sometimes a difference between the work time and the elapsed time. For example, if there is a waiting period for parts to arrive, then the elapsed time will be greater than the work time, and this elapsed time is what should be used in computing completion dates of the affected activities during the planning process. The elapsed time will always be greater than or equal to the work time. If there is significant difference, the *work time* must be used later during the cost estimating process, for it is work time that is used in computing costs, not elapsed time.

The time it takes to complete the activities should not be estimated in the same order in which the activities appear along a given path from left to right in the network. Instead, we perform the estimating in a random order. The reason for this procedure is to prevent bias that would result if the team were to estimate, in a forward sequence, all the serial activities along each given path. The people on the team would exclaim, when several of the activities along the path are estimated and the estimates are added together, that it appears much too long a schedule, and the team would probably be biased toward underestimating the activity times for the remainder of the path. Worse yet, the team might be influenced to go back and shorten the estimates for the activities all along the path. We do not need this biased, unrealistic schedule. We need a true, unbiased estimate for each and every activity. *We need an overall, accurate project schedule for the team and for higher management.*

We must estimate activity duration times in an unbiased manner.

The procedure used to prevent this kind of bias error is to use a set of random numbers (the last two digits from a five-place log table, for example) to shuffle the activities into a random order. The activities are then estimated in this new random order.

Assume that the set of random numbers found, for example, from the last two digits in a table of logarithms is the following:

16	29	03	28	38	33	32	09	23	14	42	30	17
25	26	40	41	10	04	39	12	06	31	37	24	22
13	27	18	34	05	02	15	20	21	11	43	07	19
35	45	01	44	36	08	52	46	48	51	47	50	49

We can now use this table for any project composed of fifty-two activities or less. If a project we may be working on contains more than fifty-two activities, then we merely go to a table of random numbers and select more numbers.

Example 3 from Chapter 2

CALCULATING ACTIVITY DURATIONS FOR THE PRINTER REPLACEMENT PROJECT

Applying these random numbers now to the printer replacement project first discussed in Chapter 2, Example 3, results in the assignment shown in Table 3.1. Since there are twelve activities, we need twelve random numbers. We select them from the above set of numbers. We will choose only those ranging from 01 to 12 and assign them to the activities in random order. For example, the first number we come to moving horizontally through the list is 03. The numbers 16 and 29 are outside the range 01 to 12, so we ignore them. The numbers appear in the order

03 09 10 04 12 06 05 02 11 07 01 08

and so we will select Activity 3 as the first one to be estimated, Activity 9 as the second one to be estimated, and so forth.

In Table 3.1, the numbers entered in **bold** type to the right of each activity indicate the order in which the activities are to be estimated. For example, Design Input A signal format is to be estimated first, Finalize documentation for Channel A second, and so forth. In the case of each activity, it is important to carefully consider t_m, t_o, and t_p in that order. It is important to review, as a team, the definitions of t_m, t_o, and t_p at the outset of each estimating session.

TABLE 3.1 Printer replacement activities in normal order with random numbers assigned.

ACTIVITY		ORDER IN WHICH TO ESTIMATE DURATION TIMES
a	Select printer	11
b	Assemble and install printer	8
c	Design Input A signal format	1
d	Design Input B signal format	4
e	Write main printer processing program	7
f	Design Input A interface board	6
g	Design Input B interface board	10
h	Generate databank for output form	12
i	Finalize documentation for Channel A	2
j	Finalize documentation for Channel B	3
k	Test system	9
l	Document system and system test	5

Again, we must make sure the estimates are done by those who will do the work later on. They should be edited by the best qualified people experienced in carrying out similar work, edited concurrently by the cross-functional team, and tested for accuracy and credibility.

Table 3.2 lists the activities in the specified random order along with the estimates. Having carefully estimated the activity times, we now re-order the activities so they are listed in normal order. Delete t_m, t_o, and t_p for a cleaner list. This is given in Table 3.3.

Example 4 from Chapter 2

CALCULATING ACTIVITY DURATIONS FOR THE HOME DEVELOPMENT PROJECT

Applying the random numbers now to the house building project pre-sented at the end of Chapter 2 results in the assignment shown in Table 3.4. The numbers entered in **bold** type to the right of each activity indi-cate the order in which the activities are to be estimated. For example, plumbing inspection is to be estimated first, grading the lot second, and

TABLE 3.2 Printer activities in order of random numbers with time estimates entered.					
ACTIVITY		t_o	t_m	t_p	t_e
c	Design Input A signal format	2	2	3	2
i	Finalize documentation for Channel A	3	4	4	4
j	Finalize documentation for Channel B	3	4	4	4
d	Design Input B signal format	1	1	1	1
l	Document system and system test	1	1	1	1
f	Design Input A interface board	2	5	10	5
e	Write main printer processing program	2	6	10	6
b	Assemble and install printer	5	6	7	6
k	Test system	1	2	5	2
g	Design Input B interface board	2	4	8	4
a	Select printer	2	2	4	2
h	Generate databank for output form	2	3	4	3

Table 3.3 Printer Replacement Activities in normal order with expected times.

ACTIVITY		t_e
a	Select printer	2
b	Assemble and install printer	6
c	Design Input A signal format	2
d	Design Input B signal format	1
e	Write main printer processing program	6
f	Design Input A interface board	5
g	Design Input B interface board	4
h	Generate databank for output form	3
i	Finalize documentation for Channel A	4
j	Finalize documentation for Channel B	4
k	Test system	2
l	Document system and system test	1

so forth. Again, in the case of each activity, it is important to carefully consider t_m, t_o, and t_p in that order. It is important to review, as a team, the definitions of t_m, t_o, and t_p at the outset of each estimating session.

Again, we must make sure the estimates are done by those who will do the work later on. They should be edited by the best qualified people experienced in carrying out similar work, edited concurrently by the cross-functional team, and tested for accuracy and credibility.

Table 3.5 lists the activities in the specified random order along with the estimates. Having carefully estimated the activity times, we now reorder the activities so they are listed in normal order. Delete t_m, t_o, and t_p for a cleaner list. This is given in Table 3.6.

These carefully calculated activity times are now entered on the plan network. The expected duration, in each case, is entered adjacent to the line representing the respective activity.

The time that it is expected to take to complete each activity has now been derived. The duration for each has now been estimated and this has been done for all of them. It is important to understand that these estimates are highly credible as they have been calculated by the best-qualified people in the organization. The team has worked together

Table 3.4 Home development activities in normal order with random numbers assigned.

ACTIVITY	
1–2	Identify land for home **42**
1–3	Landscape design **32**
3–4	Survey and locate foundation **3**
4–5	Excavate **19**
5–6	Underground drainage **31**
5–9	Install temporary power **22**
6–7	Pour foundation **38**
7–8	Insulate basement **45**
8–9	Backfill **8**
9–10	Plumbing service **18**
10–11	Basement plumbing **36**
9–13	Frame walls **21**
13–14	Install wall sheathing **27**
14–15	Install roof trusses **10**
11–12	Inside plumbing **33**
12–15	Plumbing inspection **1**
15–16	Roof sheathing **13**
16–17	Shingle roof **29**
17–18	Install solar system **39**
9–19	Fireplace **34**
19–20	Chimney and chimney flashing **35**
18–20	Pour basement floor **26**
20–21	Install HVAC system **9**
20–24	Windows and exterior doors **25**
24–28	Install siding **14**
21–22	Wiring **15**
22–23	Security system **28**
23–29	Wiring inspection **4**
20–25	Grade lot **2**
25–26	Driveway and sidewalk **12**
26–27	Plant shrubs **23**
27–28	Lay sod **7**
29–30	Insulate walls and ceilings **6**
30–31	Sheet rock **30**
31–32	Taping and skim coat **40**
32–34	Install cabinets **44**
32–33	Install interior doors **24**
34–35	Lay hardwood floors **5**
35–36	Sand floors **20**
36–37	Finish floors **16**
37–38	Interior finishing **17**
38–39	Light fixtures and plumbing fixtures **11**
28–39	Exterior finish **37**
39–40	Clean windows **43**
40–41	Clean property **41**

TABLE 3.5 Home development activities in order of random numbers with time estimates entered.

ACTIVITY (DAYS)		t_o	t_m	t_p	t_e (days)
12–15	Plumbing inspection	1	1	1	1
20–25	Grade lot	1	1	1	1
3–4	Survey and locate foundation	1	1	2	1
23–29	Wiring inspection	1	1	1	1
34–35	Hardwood floors	6	6	7	6
29–30	Insulate walls and ceiling	3	4	6	4
27–28	Lay sod	1	2	3	2
8–9	Backfill	1	2	3	2
20–21	Install HVAC system	5	6	7	6
14–15	Install roof trusses	1	1	2	1
38–39	Light fixtures and plumbing fixtures	3	4	4	4
25–26	Driveway and sidewalk	2	3	6	3
15–16	Roof sheathing	2	2	4	2
24–28	Install siding	4	5	8	5
21–22	Wiring	5	6	7	6
36–37	Finish floors	2	4	6	4
37–38	Interior finishing	2	2	3	2
9–10	Plumbing service	1	2	4	2
4–5	Excavate	2	2	3	2
35–36	Sand floors	1	1	1	1
9–13	Frame wall	3	3	5	3
5–9	Install temporary power	1	1	2	1
26–27	Plant shrubs	1	1	1	1
32–33	Install interior doors	3	4	4	4
20–24	Windows and exterior doors	3	4	7	4
18–20	Pour basement floor	1	1	1	1
13–14	Install wall sheathing	3	4	5	4
22–23	Security system	2	2	3	2
16–17	Shingle roof	2	2	4	2
30–31	Sheet rock	4	5	6	5
5–6	Underground drainage	2	3	4	3
2–3	Landscape design	2	2	4	2
11–12	Inside plumbing	4	5	7	5
9–19	Fireplace	3	3	4	3
19–20	Chimney and chimney flashing	3	4	7	4
10–11	Basement plumbing	3	4	5	4
28–39	Exterior finish	1	3	5	3
6–7	Pour foundation	3	3	4	3
17–18	Install solar system	2	4	6	4
31–32	Taping and skim coat	5	5	6	5
40–41	Clean property	2	6	10	6
1–2	Identify land for home	2	4	18	6
39–40	Clean windows	1	2	9	3
32–34	Install cabinets	2	3	6	3
7–8	Insulate basement	1	2	3	2

TABLE 3.6 Activities in normal order with expected times.

ACTIVITY		t_e
1–2	Identify land for home	6
2–3	Landscape design	2
3–4	Survey and locate foundation	1
4–5	Excavate	2
5–6	Underground drainage	3
5–9	Install temporary power	1
6–7	Pour foundation	3
7–8	Insulate basement	2
8–9	Backfill	2
9–10	Plumbing service	2
10–11	Basement plumbing	4
9–13	Frame walls	3
13–14	Install wall sheathing	4
14–15	Install roof trusses	1
11–12	Inside plumbing	5
12–15	Plumbing inspection	1
15–16	Roof sheathing	2
16–17	Shingle roof	2
17–18	Install solar system	4
9–19	Fireplace	3
19–20	Chimney and chimney flashing	4
18–20	Pour basement floor	1
20–21	Install HVAC system	6
20–24	Windows and exterior doors	4
24–28	Install siding	5
21–22	Wiring	6
22–23	Security system	2
23–29	Wiring Inspection	1
20–25	Grade lot	1
25–26	Driveway and sidewalk	3
26–27	Plant shrubs	1
27–28	Lay sod	2
29–30	Insulate walls and ceilings	5
30–31	Sheet rock	5
31–32	Taping and skim coat	5
32–34	Install cabinets	3
32–33	Install interior doors	4
34–35	Lay hardwood floors	6
35–36	Sand floors	1
36–37	Finish floors	4
37–38	Interior finishing	2
38–39	Light fixtures and plumbing fixtures	4
28–39	Exterior finish	3
39–40	Clean windows	3
40–41	Clean property	6

in this estimating activity. Experienced people, quite often the older, more seasoned people who have carried out similar tasks many times have been consulted. It is the best estimate possible by whatever the means or whatever the resource for the information, whether it is from people inside or outside the organization. It is obtained from the best-qualified people; therefore, it is the best information. This is an important point because the overall project estimate is only as good as the component estimates. The test is for the team to ask itself: "Could any other group, for example, a group of experienced managers (say the team's supervisors) do any better?" "Could any other group working together in estimating the activities generate a more accurate set of expected completion times for the activities?" If the cross-functional team can face each other as well as higher management and answer in the negative to these questions, then the estimates are as good as can be. Otherwise, they should be refined.

Each activity time is obtained from the best-qualified people therefore, it is the best estimate possible.

It is important to understand the reason for all this care and credibility characterizing the estimates. First, the company needs and depends on good estimates for sound business planning and execution. The company must stay ahead of competition to survive. Second, the team and its higher management is taking great care because once the estimate is submitted for approval of project funding, then no one is going to allow any rescheduling. If some higher manager attempts to shorten the schedule through suggestion, solicitation, request, demand, or coercion, it simply will not be allowed. Great care has been taken in the course of estimating the schedule, so it is the best possible and, therefore, it cannot be improved upon—by anybody.

The company needs and depends on good estimates for sound business planning and execution.

It must also be understood that these numbers that have just been calculated are still estimates and that they are merely random variables. The activities may take shorter or longer times to complete. However, they are good *estimates* and will serve a useful purpose.

They will be used to forecast the overall project completion time, and this is extremely important for the whole business organization. The production schedule, purchase of long-lead components, and the marketing schedule will all be based on this timetable. Production and marketing efforts are quite costly and the company president will want sound estimates from the cross-functional team. Following the above procedure will guarantee that the project duration estimate will be as good as possible.

It is extremely important to understand and to properly interpret for management the meaning of the estimates.

It should be pointed out that the activity duration times just calculated, when added together to obtain the estimate for the whole project, yield a 50 percent probability. This means that there is a 50 percent chance that the project will take less time to complete. It also means that *there is a 50 percent chance that the project will take appreciably longer* than the time that will be calculated by adding together the expected activity times along the longest path. This results from the mathematical process used to derive the estimates. We have done a very good job of providing the basis for an estimate of the average or mean time. This will be the mean or most likely time for the whole project. By mathematical definition, it will be the mean project time, and this carries with it a 50 percent probability implication. The reason is that we are dealing with a normal or bell-shaped probability distribution, and the mean time lies at the center of the bell.

The question that has been answered by the above calculation is this: How long is it expected to take to complete the project? The answer has been that it is expected to take x units of time to complete the project. We have calculated this estimate on a sound mathematical basis; it is valid and useful. Also, because it is based on sound mathematics, no one can do any better in calculating the expected completion time. It is based on the best available input data from the best qualified people and on statistical theory[2] that has been verified over the years in other random variable situations.

An equally important question to higher management is this: "How long will it take to complete the project with a probability of 90 percent or 95 percent?" It is easy to answer this question, also. Again, sound mathematics will be used and this will be done later in this chapter.

The point that is being made here is that in most cases, where people use network and critical path planning, only the first answer is given.

---

[2] Samuel B. Richmond, *Statistical Analysis*, Ronald Press, New York, 1964.

Management, then, is misled to think that the time that is first calculated (and matches the 50 percent probability) is the completion time with a high probability. Management misinterprets the report. They are led to believe that the reported schedule is one of high probability.

If we, the team, take a little more time and care, calculate the time for a specified higher probability, and also report that time along with a suitable description, then a degree of frustration will be avoided. It is incumbent upon the team to take this additional level of responsibility.

The calculation of project completion times matching two or three specified, higher probabilities will be covered later in this chapter.

DETERMINING THE CRITICAL PATH

Now that the schedule time has been carefully calculated for each activity, several useful tools can be derived from this information. First, the critical path (CP) will be determined, and, second, the event and activity slacks will be computed. All this data will be most useful in controlling the project, later on.

The critical path is that which, throughout the whole project, takes the longest to complete. Therefore, it is the set of activities that is the most critical to control and keep on schedule. Otherwise, if any of the critical path activities slip behind schedule, then, by definition, the whole project falls behind schedule. This puts the whole company in jeopardy (at least as far as this product is concerned) because, as stated before, the production and marketing schedules depend on the new product development schedule.

The critical path is of such importance that some organizations using network planning and scheduling use a somewhat different form of network and call it the Critical Path Method. We will continue to use the activity schedule/completion event plan network and highlight the CP on it—so as to have the best of both approaches. Furthermore, we will draw the plan network on a linear time scale. This makes it much easier to use at design reviews in conjunction with the resource allocation and cost control schedules.

We will highlight the critical path on the plan network.

Another useful tool that derives from the set of activity time estimates is the set of activity slack times. Now that it is known how long it is expected to take to complete each activity and how each is related to the critical path, the amount of slack time there is available in each activity (not on the critical path) can be calculated. This is important to know because it may be beneficial to shift resources around somewhat to gain an advantage. If, for example, the project leader can arrange with a subcontractor for the sub to come at a time more economically convenient for the sub, then it may well cost less to the project than if it were absolutely necessary for the sub to arrive on a certain, rigidly set, but less convenient date. For this reason, it is advantageous to know how much leeway there is in each activity schedule. Once the time estimates are known, it is quite easy to calculate slack times.

Activity slack will be computed on page 124. By definition, there is no slack time for any activity on the critical path.

The procedure for deriving both the critical path and the slack times will now be detailed. First, five major steps will be listed, and then each will be discussed.

1. Determine the Earliest possible Time (ET) at which each of the completion events (network circles) can occur.
2. Determine the Latest allowable Time (LT) at which each event can occur without delaying the whole project.
3. The difference, LT minus ET, gives the event slack time. *Event slack time is not the same as activity slack time.*
4. The path through all the events (network circles) that have zero event slack times is the critical path. Any ambiguities will be explained in detail later on page 108.
5. Compute the activity slack in the case of each activity. As pointed out earlier, activity slack time is different from event slack.

EARLIEST POSSIBLE EVENT TIMES

The earliest times will be computed first. They are computed by working from left to right through the network.

We assign the number 0 to the left-most event in the network, the start event. Therefore, ET = 0 for the start event. We then calculate ET for the next event to the right by considering each and every activity leading into the event. The earliest time any given activity can be completed is found by beginning with the ET for the activity's start event and adding the expected time for the activity. That is, the earliest time that the first activity can be completed is $0 +$ the t_e for the first activity. We do this for all the activities in the group leading into the event (circle) for which the new ET is being calculated. Then, take the greatest sum and this becomes the ET.

Example 1 from Chapter 2

Earliest Possible Event Times for Example 1, Buying an Automobile

In the case of the exercise first presented on page 66, there is only one activity leading into Event 2. Therefore, its ET is $0 + 1$, the activity time for the first activity, and it follows that the ET for Event 2 is 1. We find ET for Event 3 by adding the t_e for Activity 2–3, which is 5, to the ET for Event 2, which is 1. This sum is 6. That is, the ET for Event 3 is 6. The ET for Event 5 is $6 + 5 = 11$. The ET for Event 6 is $11 + 3 = 14$. The ET for Event 7 is $14 + 3 = 17$. The event time for Event 8 is the greater of $17 + 3 = 20$, calculated along Activity 7–8, and the event time for Event 3 plus the t_e for Activity 3–8, which is $6 + 12 = 18$. Therefore, the ET for Event 8 is 20, because it is the greater of the two sums.

The general formula for determining the earliest time is this:

$$\text{max: } ET = ET_{previous} + t_{e,previous}$$

considering all the paths leading into the event under consideration.

The team finds all the earliest times in this manner and enters them on the network adjacent to the corresponding events. See Figure 3.1. We will eventually enter the Earliest Time, the Latest Time, and their difference, the Event Slack time, all as a subtraction adjacent to the respective event using the format

LT = Latest allowable Time
-ET = Earliest possible Time
ES_1 = Event Slack time

FIGURE 3.1 Earliest event times for project of buying an automobile.

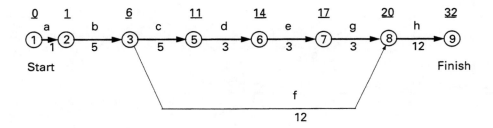

ES_1 is used here to distinguish it from ES = Earliest activity Starting time (used later in this chapter).

We will continue to work on this example on page 105.

Example 3 from Chapter 2

Earliest Possible Event Times for the Printer Replacement Project

The reader is referred to page 70, in which a typical project in a manufacturing company was introduced. The project of developing or purchasing a product and integrating it into a larger system was configured.

Having established the activity precedence and drawn the network configuration, we can now compute the earliest possible event times. Figure 2.7 is repeated below, as Figure 3.2, showing the activity times.

There is only one activity leading into Event 2. Therefore, its ET is 0+ 2, the activity time for the first activity, and it follows that the ET for Event 2 is 2. We find the ET for Event 3 by considering both Path 1–2–3 and Path 1–3. The activity time for a dummy activity is 0. Therefore, the time required for Path 1–2–3 is 2, and the time for Path 1–3 is 2. Therefore, the earliest possible time for Event 3 is 2. Similarly the earliest time for Event 4 is 2. The earliest time for Event 5 is the greater of Event 2 time plus 6 (2 + 6) or Event 3 time plus 5 (2 + 5). The greater of these two is 8.

FIGURE 3.2 Printer project network with activity times.

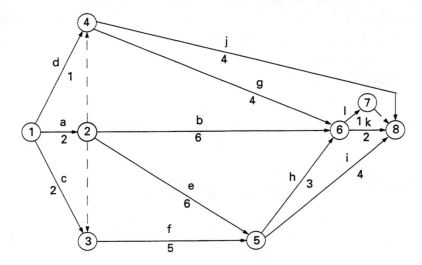

Therefore, the earliest possible time for Event 5 is 8. The earliest possible time for Event 6 is the greatest of Event 4 time plus 4 (2 + 4), or Event 2 time plus 6 (2 + 6), or Event 5 time plus 3 (8 + 3). The greatest of the three times in parentheses is 11. Therefore, the earliest possible time for Event 6 is 11, the greatest of the three sums.

This process is depicted graphically in Figure 3.3. The ETs for the whole network are given in the figure. We will continue this example on page 105.

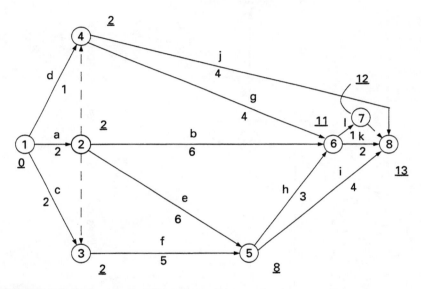

FIGURE 3.3 Earliest event times for printer project.

Example 4 from Chapter 2

Earliest Times for the Home

In the case of the home, there is only one activity leading into Event 2. Therefore, its ET is 0 + 6, the activity time for the first activity, and it follows that the ET for Event 2 is 6. We find ET for Event 3 by adding the t_e for Activity 2–3, which is 2, to the ET for Event 2, which is 6. This sum is 8. That is, the ET for Event 3 is 8. The ET for Event 4 is 8 + 1 = 9. The ET for Event 5 is 9 + 2 = 11. The ET for Event 6 is 11 + 3 = 14, the ET for Event 7 is 17, and for Event 8 is 19. The event time for Event 9 is the greater of 19 + 2 = 21, calculated along Activity 8–9, and the event time

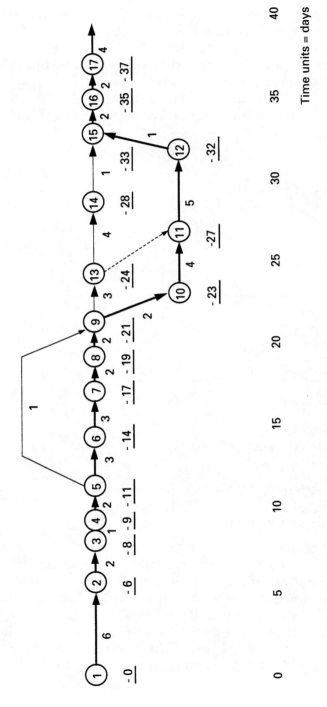

FIGURE 3.4 Graphical illustration of first few earliest times for the home development project.

for Event 5 plus the t_e for Activity 5–9, which is $11 + 1 = 12$. Therefore, the ET for Event 9 is 21, because it is the greater of the two sums.

This process is depicted graphically in Figure 3.4. The ETs for the whole first part of the network are given. Figure 3.5 lists them for the complete project. The latest allowable times will be computed next.

DETERMINING THE LATEST ALLOWABLE TIMES

The Latest allowable Times (LT) are computed by working backward from right to left through the network. The right-most event, the completion event, is given an LT identical to its ET in this book. In some books the completion event is given an LT that is less than the ET. This is done on the basis that the calculated ET is unacceptable to higher management and a shorter schedule is demanded of the team. The assumption is made in this book that the estimates are accurate—as determined by the cross-functional team working with excellent information from excellent sources, including management, in Steps 1 through 5 in the *Overall Procedure*. Therefore, no other team could possibly develop the product any faster than the schedule indicates. Therefore, it is valid and highly credible to assume that the ET derived for the completion event is, in fact, the earliest possible. Consequently, the LT must equal the ET for the completion event. Reference is made to the discussion on pages 84–89.

The assumption is made in this book that the estimates are accurate—as determined by the cross-functional team working with excellent information from excellent sources, including management, in Steps 1 through 5 in the Overall Procedure.

Therefore, no one else nor any other group could possibly develop the product any faster than the schedule indicates.

Therefore, it is valid and highly credible to assume that the ET derived for the completion event is, in fact, the earliest possible.

To determine the LT for the next to the last event, the t_e for the activity between these two events is subtracted from the completion event time. The result is the LT for this next to the last event. Where there is more than one activity emanating from an event, this backward direction

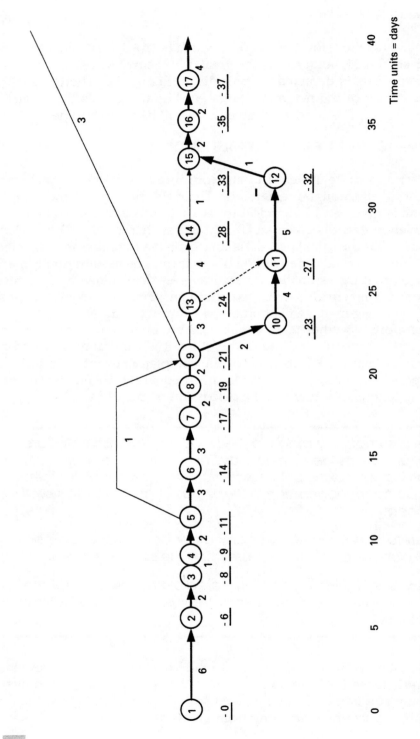

FIGURE 3.5A Graphical illustration of all the earliest times.

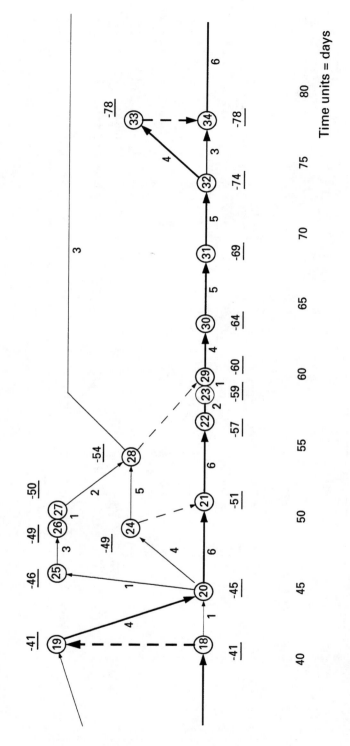

FIGURE 3.5B Graphical illustration of all the earliest times.

Time units = days

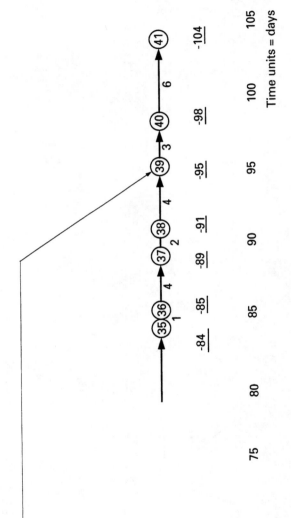

FIGURE 3.5C Graphical illustration of all the earliest times.

process must be carried out along each of the paths. The *smallest* calculated LT in the group is then assigned to the event.

Example 1 from Chapter 2

Latest Event Times for the Project of Buying an Automobile

In the project of buying an automobile, Event 9 is the completion event. The LT for Event 9 is set equal to the earliest time, 32. The LT for Event 8 is calculated by subtracting $t_e = 12$ from the LT for Event 8. The result is 20. There is only one activity emanating from Event 7; therefore, only one calculation need be considered.

Next, the LT for Event 7 will be calculated. The latest time for Event 7 is found by subtracting $t_e = 3$ from 20. Therefore, the latest time that Event 7 can occur without delaying the whole project is Day 17. The LT for Event 6 is found to be 14 by the same process. The LT for Event 5 is 11.

So far, only one calculation has been required for each event. Now, for Event 3, we must carry out the computation along two paths: along path 8–3 and along path 5–3. Along the first path, it would appear that the LT would be 20 minus 12; however, along the second path it is 11 minus 5. Therefore, the LT for Event 3 is the smaller of the two, which is 6.

This process is carried out for the whole network until all the LTs have been found. Figure 3.6 illustrates the whole project.

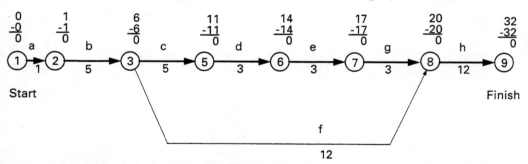

FIGURE 3.6 Earliest and latest event times and event slack time for project of buying an automobile.

Example 3 from Chapter 2

Latest Event Times for the Printer Replacement Project

Next, we will compute the latest event times for the printer replacement project that is illustrated in Figure 3.3. Event 8 is the completion event. The LT for Event 7 is calculated by subtracting $t_e = 0$ (the Dummy Activity) from the LT for Event 8, which is 13. The result is 13. The LT for Event 6 is calculated by considering the two paths that emanate from it: Path 8–7–6 and Path 8–6. Along the first path, the calculation is 13–1 = 12. Along the second path, the calculation is 13–2 = 11. The smaller of these two differences is 11; therefore the LT for Event 6 is 11.

For Event 5, we must carry out the computation along two paths: along path 8–5 and along path 8–6–5. Along the first path, it would appear that the LT would be 13 minus 4; however, along the second path it is 11 minus 3. Therefore, the LT for Event 5 is the smaller of the two, which is 8.

This process is carried out for the whole network until all the LTs have been found. Figure 3.7 illustrates the whole project.

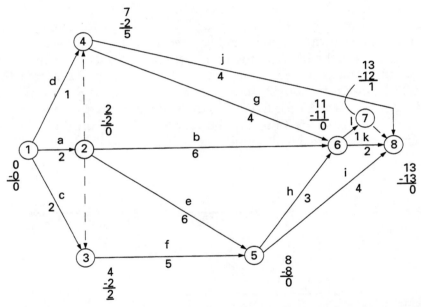

FIGURE 3.7 Earliest and latest event times and event slack for printer project.

Example 4 from Chapter 2

Latest Event Times for the Home

In the home-building project, which is illustrated in Figure 3.8, Event 41 is the completion event. The LT for Event 40 is calculated by subtracting $t_e = 6$ from the LT for Event 41, which is 104. The result is 98. There is

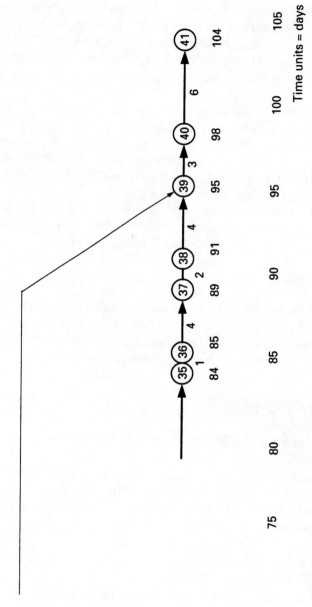

FIGURE 3.8 Latest times for the last few events.

only one activity emanating from Event 40; therefore, only one calculation need be considered.

Next, the LT for Event 39 will be calculated. The latest time for Event 39 is found by subtracting $t_e = 3$ from 98. Therefore, the latest time that Event 39 can occur without delaying the whole project is Day 95. The LT for Event 38 is found to be 91 by the same process. The LTs for Events 37, 36, 35, 34, and 33 are 89, 85, 84, 78, and 78, respectively. See Figures 3.8 and 3.9.

So far, only one calculation has been required for each event. Now, for Event 32, we must carry out the computation along two paths: along path 34–32 and along path 34–33–32. Along the first path, it would appear that the LT would be 78 minus 3; however, along the second path it is 78 minus 4 (the dummy path has $t_e = 0$). Therefore, the LT for Event 32 is the smaller of the two, which is 74.

See Figure 3.8 for an illustration of the latest times for the last few events. This process is carried out for the whole network until all the LTs have been found. Figure 3.9 illustrates the whole project.

EVENT SLACK TIME

The event slack times are found by merely subtracting the earliest time from the latest time in each case. See the figures for an illustration. As stated earlier, the earliest time, the latest time, and their difference, the event slack, are all entered in a subtraction format adjacent to the respective event.

THE CRITICAL PATH DEFINED

The path with the longest total time through the network from start to finish is called the critical path.

The Critical Path (CP) is defined as the longest path through the network that passes through all the events with zero slack times. In somewhat more detail, each and every activity on the CP must meet the following requirements:

1. The event slack of the start event, that is, the event immediately preceding the activity, must be zero.
2. The event slack of the finish event must be zero.
3. The difference between the finish event time and the start event time for the activity must equal the activity duration time.

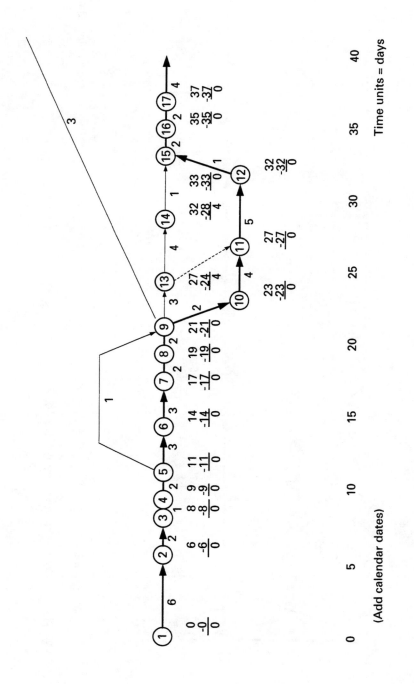

FIGURE 3.9A Plan Network for development of home.

Time units = days

(Add calendar dates)

109

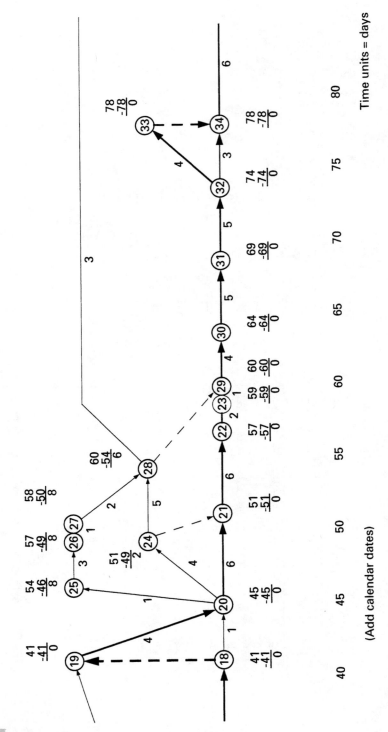

FIGURE 3.9B Plan Network for development of home.

Time units = days

(Add calendar dates)

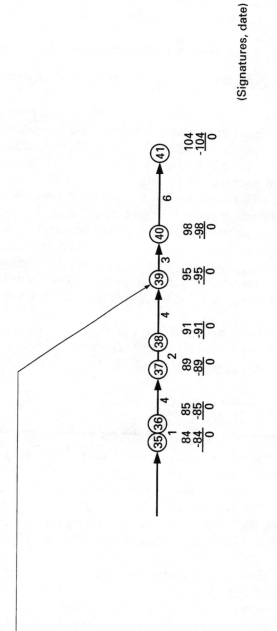

FIGURE 3.9C Plan Network for development of home.

Time units = days

(Add calendar dates)

(Signatures, date)

le, in the home development project, the critical path is 32–33–34 instead of through Events 32–34 because of ve.

path for the printer project is through Events 1–2–5–6–8.

PATHS AND OTHER ISSUES IN NETWORK PLANNING

So far we have considered the critical path, only, in our network planning. We must ask ourselves, "What about other activity sequences that are running in parallel with the so-called critical path?" "What about the resources required to carry all these activities out?" "What are the costs?" How do near-critical paths impact the scheduled project completion date?

These areas and others will be covered in the next sections. We wish to cover next the matter of probability of project completion by certain specified dates, or more to the point, we wish to compute the completion date(s) matching a specified probability of success in meeting that date. The expected completion time, t_E, is merely the sum of all the individual critical path activity times. That is, t_E = the sum of all the individual expected times, t_e. This overall expected completion time carries a probability of success of 50 percent.

An estimate of completion time matching a prespecified probability of 90 or 95 percent is at least equally interesting. The next section covers this higher probability estimating procedure.

THE MORE MEANINGFUL PROJECT SCHEDULE

Previously, we pointed out that the sum of all the expected activity times, t_e in each case, is the expected time for the project. We pointed out that this expected time is to be interpreted as carrying with it a probability of 50 percent. That is, there is a 50 percent probability that we will complete the project in less time, and *there is a 50 percent probability that the project will take significantly more time.* This can be proven mathematically, with sound results.

The natural question that we, who are all managers (of our own time, our performance, and, therefore, our destiny), ask is: How long will it take us to complete the project with a probability of 90 percent or 95 percent? We understand that we cannot deal with 100 percent certainty because, in addition to our intuitive feel, the mathematics teach us that the dynamics and the phenomenon of project activities follow the normal, bell-shaped curve, and that to get to a probability of 100 percent would require an infinitely long time. This is illustrated in Figure 3.10, which is the normal probability density function. The ends of the curve never quite reach the zero probability horizontal axis.

Suppose a large number of identical projects were carried out by different teams of skilled people. Figure 3.10 shows that most would complete their respective projects at or around the expected time, t_E. Some teams would take less time and some would take more. Because half the area under the curve lies to the left of t_E and half to the right, the implication is that half the teams would take less time than t_E and half would take more. However, no team would take an infinitely long time to complete its project. Therefore, the normal curve, matching these dynamics, illustrates this fact. In order for the total area under the curve, which represents a probability of 100 percent, to be taken into account, this must cover even that team which would take an infinitely long time. Since no team would take this long, we understand that we are always dealing with probabilities of less than 100 percent.

The important, distinguishing factor in determining the expected project completion time for a prespecified probability is the standard deviation of the normal curve. Suppose a company officer were to ask, "Given a degree of certainty of, say, 95 percent, how long do you think it will take to complete the project?" We would like to be able to answer, with all the confidence of aspiring, well-informed employees: We are 95 percent sure that we can complete the project in ten months. Then we would proceed to give a brief report on how we arrived at the answer,

FIGURE 3.10 Normal probability functions.
  (a) probability density function
  (b) probability distribution function

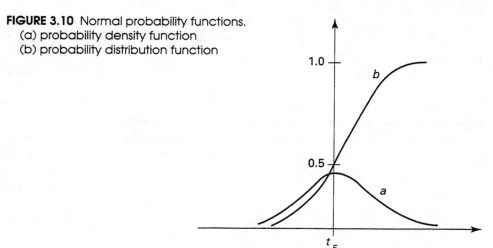

and the officers can move forward with company planning. The company engineering, marketing, manufacturing, and financial planning can all be tied together.

The purpose of this section is to provide the details of just how this is done—with great credibility, that is, based on sound mathematics.

First, we will discuss the variation in activity completion times, given the same activity but imagining that it is carried out by different groups of similarly skilled people. This variation, in statistics, is called *variance.* Its square root is called the *standard deviation,* an equally interesting term in statistics.

The standard deviations of the statistics of expected activity times determine the standard deviation of the overall project normal curve. (The reader is not yet overwhelmed, at least not by this statement.)

In this section, we will learn how to estimate the completion time matching a 95 percent probability of success.

If there is little variation in the deviation times, then the normal curve is tall and narrow. This would be the case where activities are well known and more predictable. If there is wide variation, in a different type of project, perhaps one where there is more research required, then the normal curve is shorter and wider. We know from mathematics that, in any case, the area under the curve is always the same, regardless of the standard deviation, and that it has been normalized to unity. The area is always unity, regardless of the deviation in statistics. This is equivalent to stating that every one of the teams, given an infinite amount of time, will complete the project with a 100 percent certainty.

The time required to complete a project can be illustrated by Figure 3.11.

Suppose that the standard deviation of a particular type of project is very small. Imagine that we are setting out to develop a new product that is well known, requires little research and carries few unknowns. Suppose, for example, that we want to release a new product line where the only difference is a whole new color scheme. We know the color exactly, the paints are all in stock, the people are all available immediately, and the print shop can produce new colored sales brochures immediately. There is little chance that the computer product codes will take longer than the other activities, and little chance for wide deviation in carrying out a plan of this type.

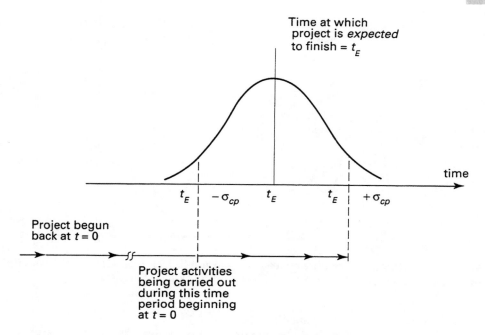

FIGURE 3.11 Normal curve used to compute project completion date.

You may wonder whether it is even worth while to carry out extensive planning on such a straight forward project. For purposes of illustration, though, planning would produce a normal curve that appears as a very narrow curve in Figure 3.12.

Projects that involve significant research as part of the development program have relatively wide standard deviations.

On the other hand, let us consider a project with many unknowns—and they do exist. Sometimes we call them the *known unknowns*. Projects that involve significant research as a part of the research and development program are of this nature. There are many such programs. These projects, too, must be estimated. This is the type of project that the cynical comment on. The statement is sometimes heard, You cannot schedule invention. We are not attempting to schedule invention. We are scheduling a research and development project. And we who are the individuals of the cross-functional team are the best qualified to estimate the project

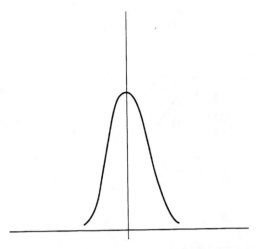

FIGURE 3.12 Normal function with relatively small variance.

FIGURE 3.13 Normal function with relatively large variance.

completion date. We will assign the research activities, with the known unknowns, relatively wide deviation. The optimistic and pessimistic times will differ widely. It may even be prudent to assume that, pessimistically, three or four or more iterations will have to be taken before the research answers to a particular technical problem will be found. Even the most likely time, t_m, may be based on more than one iteration.

Figure 3.13 illustrates a project with a relatively wide variance.

Whatever the case may be, the cross-functional team is the organization that makes the most intelligent estimates, calculates the project completion time, and reports the findings to the company officers. The company now can move forward with the overall integrated engineering, marketing, manufacturing, and financial business plan.

It should be reemphasized, here, that the cross-functional team must provide the project completion time. If anyone else in the company attempts to preempt or supplant the estimate, it will not be nearly as valid.

Let us proceed, now, with the estimating process that provides the project completion time that matches a probability of success of 95 percent. The question that must be answered is, What is the completion time for the given technical project, that matches a 95 percent chance of success of meeting this completion time? We will now detail how this is done with great care. It will be based on sound, easily understood mathematics and, therefore, the process is highly credible.

This estimate is quite easily made. It is important that the team be provided with a project management tool for planning, scheduling, and control that is easily formulated and easy to use. The procedure is the following:

1. Review each of the activity estimates to confirm that each estimate was made by the best qualified individual(s); that there has been input, where applicable, by manufacturing engineering, service, maintenance, and quality engineering personnel, as well as product design personnel; and that the whole procedure has been edited and validated by the whole cross-functional team. *The essence of concurrent engineering is team input and agreement.*

2. Highlight the activities on the critical path. Calculate the variance of each activity on the CP. This is done by calculating for each activity:

   variance = $\sigma^2 = ((t_p - t_o)/6)^2$, σ = standard deviation.

3. Calculate the standard deviation for the whole CP. This is done by adding all the individual variances together to obtain the sum of all the CP activity variances. Then the square root of the CP variance is the standard deviation, σ_{cp} of the CP. The σ_{cp} gives the relative width of the normal curve for the critical path of the technical project under study.

4. Apply well-known statistics theory to the problem at hand. For example, one standard deviation covers 34 percent of the area under the normal curve on either side of the mean. The mean matches the expected time, t_E. The deviation of $\pm\sigma_{cp}$ covers 68 percent of the area under the normal curve. That is, there is a 68 percent chance of completing the project between $t_E - \sigma_{cp}$ days and $t_E + \sigma_{cp}$ days. See Figure 3.11.

To complete the project by a specified number of days, t_{ST}, with a 95 percent chance of success implies that the area under the normal curve must be 50 percent + 45 percent. Therefore, we must find a time, t_{ST}, to the right of t_E that corresponds to an area of 0.45. This is done by using a table of areas derived from the standard normal curve. Normal deviates, which are factors which relate standard deviation to the corresponding area under the standardized normal probability distribution curve, are given in Table 3.7. See Figure 3.10, where Curve *b* is the integral of Curve *a*.

In order to find the probability, one always equates the area under the standard normal, bell-shaped curve to the probability sought. That is, if for example, one wants to find the probability of completing a project by a certain specified schedule time, t_{ST}, then the steps are the following:

1. Compute $Z = (t_{ST} - t_E)/\sigma$.

    Say $t_E = 40$, $t_{ST} = 45$, and $\sigma = 3.33$.

    Then $Z = 1.5$.

2. The probability is given in Table 3.7 as 93.3 percent.

Use Table 3.7 to find the probability matching a particular Z value. There are four sets of double rows of data. The top row in each set contains the Z values. The bottom row contains the corresponding probabilities. For example, if $Z = -1.4$, then the probability $= .081$ or 8.1 percent. If $Z = 0.7$, then the probability $= .758$ or 75.8 percent.

For values of Z less than -2.0, the probability is essentially zero. For values of $Z = +2.0$ or greater, the probability is essentially 100%.

5. Then the equation relating the normal deviate factor to the project completion time and the project standard deviation is:

    $$Z_{cp} = (t_{ST} - t_E)/\sigma_{cp}$$

    For example, given that the area is $.50 + .45 = 0.95$, then the corresponding Z factor is $Z = 1.65$.

6. Find t_{ST} by manipulating the equation in Step 5 to read:

    $$t_{ST} = t_E + Z_{cp}\sigma$$

TABLE 3.7 *Z* values and corresponding probabilities of success.

-2.0	-1.9	-1.8	-1.7	-1.6	-1.5	-1.4	-1.3	-1.2	-1.1
.023	.029	.036	.045	.055	.067	.081	.097	.115	.136
-1.0	-0.9	-0.8	-0.7	-0.6	-0.5	-0.4	-0.3	-0.2	-0.1
.159	.184	.212	.242	.274	.309	.345	.382	.421	.460
0.0	0.1	0.2	0.3	0.4	0.5	0.6	0.7	0.8	0.9
.500	.540	.579	.618	.655	.691	.726	.758	.788	.816
1.0	1.1	1.2	1.3	1.4	1.5	1.6	1.7	1.8	1.9
.841	.864	.885	.903	.919	.933	.945	.955	.964	.971

In the example, $t_{ST} = t_E + 1.65\sigma$, which gives the completion time with 95 percent chance of success.

The above procedure of six steps results in a fairly accurate estimate of the project completion time. It is a good first approximation. If there are paths in parallel with the CP, or parallel for part of the way, and if they have expected completion times that are not significantly less than the CP expected time, then the probability of completing the overall project must undergo one additional step. This step will be covered in the next section. If the CP is the longest path, by far, then the above steps are all that are necessary.

In the home development project, the activities on the CP: $1-2-3-4-5-6-7-8-9-10-11-12-15-16-17-18-19-20-21-22$ $-23-29-30-31-32-33-34-35-36-37-38-39-40-41$ are given in Table 3.8. The variance is listed for each of these activities, and the sum of the variances is given at the bottom of the table.

The standard deviation is calculated to be

$$\sigma_{cp} = \sqrt{\text{var}} = 3.745 \text{ days.}$$

Therefore, $t_{ST} = t_E + Z\sigma$, $Z = 1.65$ for a 95 percent chance of completing by the scheduled time, ST. Therefore, in this example, $t_{ST} = $ Day 110.

Figure 3.14 illustrates the 95 percent case and a 15 percent case—for comparison.

The procedure carried out above was a step-by-step process for calculating a project completion time carrying a 95 percent chance of success. The same process could be carried out for a 90 percent or a 98 percent probability, for example.

We must remember that there should be at least thirty activities along the critical path to invoke the Central Limit Theorem, a requirement for using the standard normal curve.

PARALLEL PATHS

In the case where there are parallel paths, in addition to the critical path, scheduled to finish at approximately the same time as the CP, a somewhat more complex calculation is required. It sometimes happens that one or more paths have schedules that are not significantly less than the critical path. In that case, we use the mathematics of joint probability theory to calculate the completion date of the project. It is not difficult, yet demands

TABLE 3.8 Critical activities in normal order with variances.

ACTIVITY		t_e	VARIANCE
1–2	Identify land for home	6	7.111
2–3	Landscape design	2	0.111
3–4	Survey and locate foundation	1	0.028
4–5	Excavate	2	0.028
5–6	Underground drainage	3	0.111
6–7	Pour foundation	3	0.028
7–8	Insulate basement	2	0.028
8–9	Backfill	2	0.111
9–10	Plumbing service	2	0.250
10–11	Basement plumbing	4	0.111
11–12	Inside plumbing	5	0.250
12–15	Plumbing inspection	1	0.000
15–16	Roof sheathing	2	0.111
16–17	Shingle roof	2	0.111
17–18	Install solar system	4	0.444
19–20	Chimney and chimney flashing	4	0.444
20–21	Install HVAC system	6	0.111
21–22	Wiring	6	0.111
22–23	Security system	2	0.028
23–29	Wiring Inspection	1	0.000
29–30	Insulate walls and ceilings	4	0.250
30–31	Sheet rock	5	0.111
31–32	Taping and skim coat	5	0.028
32–33	Install interior doors	4	0.028
34–35	Lay hardwood floors	6	0.028
35–36	Sand floors	1	0.000
36–37	Finish floors	4	0.444
37–38	Interior finishing	2	0.028
38–39	Light fixtures and plumbing fixtures	4	0.028
39–40	Clean windows	3	1.778
40–41	Clean property	6	<u>1.778</u>
			14.028

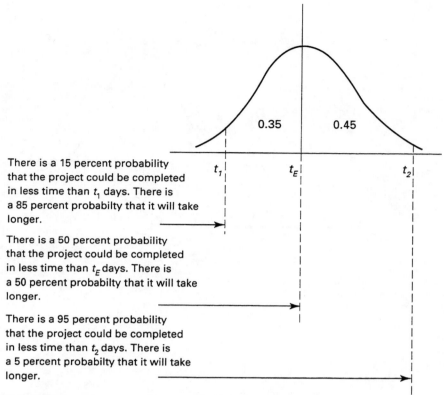

There is a 15 percent probability that the project could be completed in less time than t_1 days. There is a 85 percent probabilty that it will take longer.

There is a 50 percent probability that the project could be completed in less time than t_E days. There is a 50 percent probabilty that it will take longer.

There is a 95 percent probability that the project could be completed in less time than t_2 days. There is a 5 percent probabilty that it will take longer.

FIGURE 3.14 Calculation of project completion time.

a necessary, somewhat more comprehensive study of the path schedules. Let us illustrate the procedure by considering two paths, Path 1 and Path 2, as in Figure 3.15. Say Path 1 has been identified as the critical path. The procedure is to do the following:

(a) Draw the Path 1 normal curve in alignment with Path 1 ,with, of course, t_{E_1} located at the Path 1 expected completion date.

(b) Draw the Path 2 normal curve in alignment with Path 2, with, as before, t_{E_2} located at the Path 2 expected completion date. The Path 2 normal curve is found by considering all activities on Path 2, the corresponding variances, and the resulting standard deviation.

Statistics theory teaches us that the probability of completing the project by a specified date is the product of the two individual probabilities for Path 1 and Path 2, *if the two sequences are statistically independent.* That is, if carrying out the activities on one path does not affect the schedule of activities on the other path, then the two are independent. We can then use the product of probabilities.

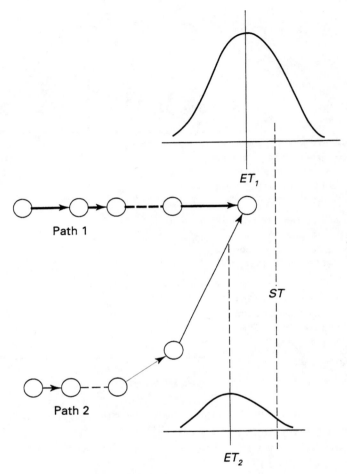

Path 1

Path 2

ET_1

ST

ET_2

FIGURE 3.15 Calculation of joint probability.

Because we were careful to properly allocate resources, we know that the two paths are independent. We were quite careful to be realistic. When it was thought that we could shorten the overall time table by scheduling more parallel paths, we did so. But remember that we were realistic; we did not do this blindly. We, in fact, scheduled some activities sequentially instead of in parallel because of resource limitations.

We can now use this fact that the paths are independent in employing the theory of statistical independence.

(c) Compute $(t_{ST} - t_{E_1})/\sigma_1 = Z_1 \Rightarrow P_1$. For example,

if $t_{ST} = $ Day 85, $t_{E_1} = $ Day 76 and $\sigma_1 = 10$, then
$Z_1 = (85 - 76)/10 = 0.9$, which implies that
$P_1 = 50 + 0.32 = 0.82$.

(d) Compute $(t_{ST} - t_{E_2}/\sigma_2 = Z_2 \Rightarrow P_2$. For example,

if $t_{E_2} = $ Day 66, and $\sigma_2 = 15$, then
$Z_2 = (85 - 66)/15 = 1.267$, which implies that $P_2 = 0.897$.

(e) Compute the probability of completing both paths by Day 85
$P = P_1 P_2 = 0.732$

It may be that we are not satisfied with this result. For example, we may be attempting to determine a completion date matching a greater overall project probability, say 95 percent. At this point we would write a small simulation program and execute it. The rudiments of the procedure, though, are illustrated in the continuing procedure.

(f) Adjust t_{ST} to Day 90, say, and repeat the process. We know we must go to a longer time than Day 85 because we want to increase the probability from 73 percent to something greater.

$Z_1 = (90 - 76)/10 = 1.4$, which implies that

$P_1 = 0.5 + 0.419 = 0.919 = 92\%$.

We see that Day 90 is still not enough because we are at 92 percent and multiplying it by the second probability will only decrease it.

(g) Adjust t_{ST} to Day 95

$Z_1 = (95 - 76)/10 = 1.9$, which implies that

$P_1 = 0.5 + 0.4713 = .9713$

(h) $Z_2 = (95 - 66)/15 = 1.933$, which implies that

$P_2 = 0.5 + 0.47436 = 0.9736$

(i) $P = P_1 P_2 = 0.9713 \times 0.9736 = 0.95 = 95\%$.

The conclusion, therefore, is that in order for us to promise higher management a completion date with a 95 percent probability of completing the project by that date, we must state that the completion date will be Day 95. The fact that the completion date, Day 95, is equal to the probability, in this case, is purely coincidental.

This procedure will work for any specified probability.

ACTIVITY SLACK

We pointed out in the section in Chapter 3 dealing with the critical path that it would be important for us to address the issue of activity slack. We state, again, that activity slack is different from event slack. Event slack is the difference between the Latest allowable event Time (LT) and the Earliest event Time (ET), and their calculations are used for determining the critical path. These calculations will also be used, now, for computing activity slack.

Activity slack applies only to activities not on the critical path. By definition, activities on the CP have no slack time, and each must begin at the preceding event time and finish at the succeeding event time. There is no leeway.

Activity slack is the amount of leeway we have in scheduling an activity. It is important for us to know how much slack we have in scheduling an activity, for it may greatly benefit us to schedule an activity for one period rather than another. For example, it may be much easier for a subcontractor to construct the fireplace on Days 36, 37, and 38 than at any other time. That period happens to be between jobs for him, and he has someone available then. If we were to demand that he construct the fireplace on Days 22, 23, and 24, instead, he may have to provide someone else at a much higher cost. It is, therefore, beneficial to know how much slack an activity has.

Activity slack is calculated by observing event times and expected activity durations. We see in Figure 3.16 that the earliest Activity 9–19 can finish, EF, is Day 21 plus the expected duration for the fireplace activity or 21 + 3. That is, the earliest Activity 9–19 can finish is on Day 24.

$$EF = Day\ 21 + t_{e,\ fireplace} = Day\ 21 + 3 = Day\ 24$$

We also see that the latest Activity 9–19 can start, LS, without impacting the overall project schedule, is Day 41 minus the expected duration for the fireplace activity or 41–3. That is, the latest Activity 9–19 can start is Day 38.

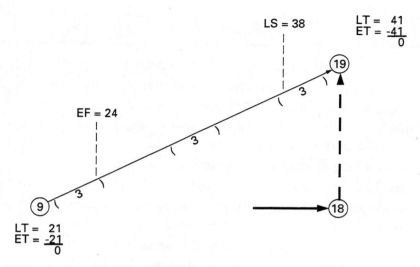

FIGURE 3.16 Calculation of activity slack.

$$LS = Day\ 41 - t_{e,\ fireplace} = Day\ 41 - 3 = Day\ 38$$

The activity slack is calculated as the LT for the succeeding event minus EF or, alternatively, LS minus ET for the preceding event. Both these calculations are illustrated in Figure 3.16. It can also be seen that the third way of defining slack, or leeway, is to begin with the spread between LT and ET (which, in our example of the fireplace, is Day 41 minus Day 21, which is twenty days) and then subtract out the activity duration, three days. The result is eighteen days of leeway for scheduling the fireplace.

$$Activity\ Slack = LT_{succeeding\ event} - EF$$

or

$$Activity\ Slack = LS - ET_{preceding\ event}$$

or

$$Activity\ Slack = LT_{succeeding\ event} - ET_{preceding\ event} - the\ expected\ time$$
for the fireplace.

We can calculate the activity slack for each of the activities not on the critical path in this manner.

PROBLEMS

3.1 Which of the following are true? The plan network is used to:

(a) formulate the manner in which we will carry out the project.

(b) familiarize everyone on the cross-functional team with the project.

(c) provide a management tool for assessing the progress and status of the project and for controlling the project.

(d) provide interim reports to higher management.

(e) provide a substitute for cost control.

(f) provide an input for generating the resource allocation.

(g) provide a substitute for Gantt charts.

(h) provide a substitute for resource allocation.

(i) all of the above.

3.2 Which of the seven major elements of the *Overall Procedure* should be *completed* prior to the rough draft of the plan network?

3.3 Which of the seven major elements of the *Overall Procedure* should be *completed* prior to finalization of the plan network?

3.4 How does the most likely time, t_m, differ from the expected time?

3.5 The meaning of t_o is that it is the time that yields the probability of 5 percent that it will take less time than t_o to complete the activity. T or F?

3.6 The meaning of t_p is that it is the time that yields the probability of 5 percent that it will take more time than t_p to complete the activity. T or F?

3.7 Why is t_m given four times as much weight in computing t_e than either t_o or t_p?

3.8 Why do we estimate the activity duration times in random order?

3.9 The calculation of all the activity times results in the expected *project* completion time. Why is this completion time only 50 percent probable?

3.10 Using the activity durations for Figure 2.8 given in Problem 2.10,

(a) draw the plan network to scale.

(b) enter the ET adjacent to each event using our standard format.

(c) give the earliest project completion time.

3.11 For the project in Problem 3.10,

(a) enter the LT adjacent to each event using our standard format.

(b) enter the event slack time.

(c) find and indicate the critical path.

3.12 Given the following information:

ACTIVITY	t_e	IMMEDIATELY PRECEDING ACTIVITY
a	2	—
b	10	a
c	10	b
d	6	c
e	6	d
f	14	b
g	6	e
h	14	f, g

(a) Draw a rough plan network.

(b) Redraw the network on a linear time scale.

(c) Enter the LT, ET, and the event slack adjacent to each event.

(d) Find the CP and indicate it.

3.13 Repeat Problem 3.12 with the activity times for b and c equal to 5 each, and the activity time for f equal to 30.

3.14 Given the following information:

ACTIVITY	t_e	PRECEDING ACTIVITY
a	4	—
b	3	a
c	5	b
d	5	b
e	6	b
f	18	c, d, e
g	9	d
h	15	f
i	2	g, h
j	3	i

(a) Draw a rough, unscaled plan network.

(b) Draw the network on a linear time scale.

(c) Enter the LT, ET, and event slack adjacent to each event

(d) Find the CP and indicate it.

3.15 Given the following information for Figure 2.8:

ACTIVITY	t_o	t_m	t_p
1–2	4	5	6
2–3	1	3	5
3–5, 5–6, 6–7	2	3	4
3–8	6	12	18
7–8, 8–9	1	3	5

(a) Calculate t_e for the CP.

(b) Calculate the project completion time that has a 95 percent probability. Assume all activities have a normal distribution.

3.16 Given the project represented by the network below,

(a) Find t_E for the CP.

(b) Find the joint probability of Event 5 occurring by the t_E found in (a).

(c) Find the project completion time corresponding to a joint probability of 95 percent.

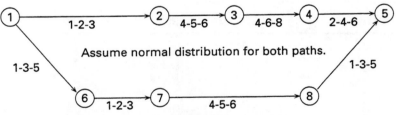

FIGURE P3.16 Joint probability illustration for Problem 3.16

3.17 Find the slack for each activity in Problem 3.14.

4 CONCURRENT SCHEDULING—RESOURCE ALLOCATION

INTRODUCTION

We have now carried out steps 1 through 4 in the *Overall Procedure*. It is repeated on pages 130–131, for easy reference, with these steps printed in **bold** type. The plan network is only a draft at this point because we have not yet checked resources. They may be available when called for; they may not be. The reader is referred to Chapter 2, including Figure 2.1.

The next step is to arrange for the resources to be available when the schedule calls for them, if possible. It is important to do this *and to gain the respective supervisor's concurrence in each case* during the planning and scheduling phases. Some supervisors will be from other departments, where talent is drawn from, in a company with a matrix organization.

The task list in Table 2.1 indicates the various types of talents required. Table 3.1 is a more complete list. Each task requires a distinct talent, and the project leader, with the assistance of the rest of the cross-functional team, identifies the specific type of individual for each activity. This work breakdown structure can be illustrated in the form of a vertical bar chart or histogram. This will be done later in this section.

First, it will be necessary to validate the resource allocation by gaining concurrence from all the cross-functional team members and from the managers of the departments from which the various talented individuals will be drawn. There will be a natural resistance against providing resources to a project outside any given department. So it was, so it is, and so it always will be. However, in the final analysis and negotiation, it will be agreed by all that resources must be provided. Management knows that matrix resourcing is best for all, and concurrence can be gained when the reasons for providing resources from Department A to Department B are well explained. Also, someday, resources will be required from B to A, and managers who have been cooperative in the past will benefit in the future. Reasonable people understand this.

OVERALL PROCEDURE

1. Clarify the *FUNCTIONAL SPECIFICATIONS*. Study the elements of planning to make sure you and your teammates have properly addressed all aspects of planning. Clarify the company objectives and the marketing objectives of the new product development. Study competition. Determine customers' needs. Write product description. Write final functional specifications. Document agreement with management team by everyone on the team signing the final specifications.

2. Detail the product specifications. Identify and establish the cross-functional team. Study competition. Study the functional specification. Select the design concept. Detail all aspects of the physical product. Rely on experts, Write a *DETAILED PRODUCT DEFINITION*. Document agreement within the cross-functional team by having all members of the team sign the detailed product definition.

3. Describe the project. Elements 1 and 2, above, describe the product. This element, the *PROJECT DESCRIPTION*, describes the project of developing the product. List all the tasks composing the project—list approximately forty or fifty tasks. Rely on experts. Write a detailed project description. Document agreement within the team in the same manner as above.

4. Generate the *PLAN NETWORK*. List all the tasks/activities that compose your project. Prioritize them. Indicate after each activity the one or more other activities that immediately precede the one under study. Draw a rough network to illustrate the correct configuration. Rely on experts. Study resources. All members of the cross-functional team must sign the plan network.

5. Validate the **RESOURCE ALLOCATION.** Draw the work breakdown structure and assign the different talents to corresponding tasks. Obtain approval from the source managers for scheduled allocations. Modify the plan network configuration if necessary. *Redraw the plan network on a linear time scale.* Identify the critical path. All members of the cross-functional team must sign the revised plan network.

   Draw the resource allocation in the form of a histogram/vertical bar chart on the same time scale as used for the plan network. All members of the cross-functional team must sign the resource chart.

6. Generate the **COST SCHEDULE.** Draw a cost curve as a function of time using the same time scale as used for the plan network and the resource histogram. All members of the cross-functional team must sign the cost curve.

7. **REPORT** the plan to management and request funding. Because experts were relied upon for the above planning, the best possible plan has been put in place. The project leader has the responsibility of presenting the plan for approval.

RESOURCE ALLOCATION

The process of resource allocation is as follows:

1. Begin with the plan network that was roughed out in Chapter 3. We recognize that this is based on the *expected* schedule, which is 50 percent probable. Later on, in on pages 153–155, we will adjust the allocation to fit the 95 percent schedule.

2. Make sure the *Overall Procedure* is followed.

3. *Block out* the resource allocation for the critical activities. So far this is what is required, not necessarily what is available at the scheduled times. *Blocking out* means to graphically illustrate the required resource in the form of a bar chart as in Figure 4.1.

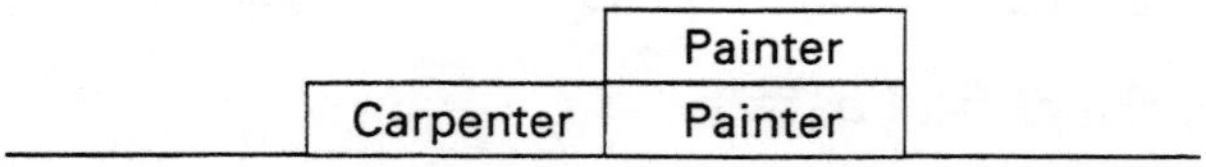

	Painter	
Carpenter	Painter	

FIGURE 4.1 Section of a resource allocation chart.

4. Block out the resource allocation for the other activities.

5. Check availability of all resources required for all activities; this is a very important process. Check availability for each and every activity: identify each person, by name, and note whether that person is from the project manager's department or from some other department. If the latter, meet with the person and his or her manager. Get both to sign availability when the schedule calls for that corresponding talent. If they are unwilling to commit the resource for that period, then determine the best time in the future that they can commit to and get them to sign for that period.

This process could well result in having to change the projected schedule. Whatever it leads to, it is important to

1. Determine with all responsible parties when availability can be guaranteed. Keep overall project as short as possible.

2. Have the persons who will carry out the activities and their respective managers sign commitments that, in fact, these persons will be available for the corresponding activities when the time comes. I have found in working with many companies that this is a process not often carried out, and that it is the reason for schedule slippage later on. On the other hand, I have found that when this process is carried out, then camaraderie, concurrence, and success are provided.

   Enter the committed individuals' names at each and every corresponding activity in the plan network. See Figures 4.11 and 4.12 at the end of this chapter.

3. Make the plan network agree with the guaranteed resource allocations, by redrawing it, if necessary. Get signed commitment from participants, and enter their names on the respective activities on the network.

   Final editions of the plan networks appear at the end of this chapter with the names of individuals entered.

4. For this guaranteed resource allocation and the corresponding plan network, redraw the resource allocation chart in the form of a block histogram for illustration and easy communication with all.

   Do this for the critical path.

   Do this on a second level for all the other activities. See Figures 4.2 and 4.3.

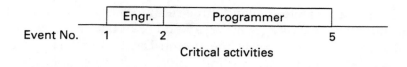

Critical activities

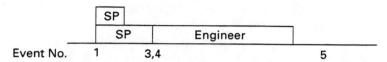

Non–critical activities

FIGURE 4.2 Section of resource allocation chart for critical and non-critical activities for printer project.

> *We now have a relatively firm plan showing the project in much "harder" form.*

We now have a relatively firm plan showing the project in much "harder" form. Where before we had a network that was quite useful, it was "soft" because resources had not yet been checked. It was worth doing, though. We had to start with something, and it did illustrate the project and gave us a place to start in checking resources.

Let us review, now, the reasons and the motivation for being very careful when we allocate resources.

Each activity, by definition, requires special resources in order to carry the activity out to completion. In the example of the home-building

FIGURE 4.3 Section of resource allocation chart for critical and non-critical activities for home development project.

project, we must arrange for landscape design talent to be available to carry out the landscape design on Days 7 and 8, and have the services of a surveyor on Day 9. To excavate the site on Days 10 and 11, we should obtain a commitment from other individuals. We must obtain commitments from all the various people with the appropriate talents to carry out the work on the days designated. We must gain their concurrence to do so. This is so serious that we will even ask them to sign an agreement to this effect, and, in fact, give them each a copy of the plan network so that they can view the big picture, see their names on the network adjacent to their activities and feel a part of the project. We are asking them to "buy-in."

The problem that many organizations have with schedule slippage is that they have failed to obtain serious buy-in from the resource managers at the outset of project planning. The individual responsible for the product development assumes that the required resources will automatically be available whenever they are needed. Many times a promise is obtained at a meeting or over the telephone. This promise is not recorded by the individual making the promise. He or she was sincere at the time, but the promise was not recorded and therefore not remembered. Consequently, when the talent is required at some point in the future, it is not available.

An important reason for making resource arrangements early on is that often they cannot be promised for the dates first indicated by the plan network. Resources are very valuable and, therefore, will probably be committed to other tasks if we wait until just before they are needed. If we check availability early on, then the plan network can be adjusted to accommodate whatever conditions exist before we submit the plan for approval. It will probably be the case that a minimum of adjustments will have to be made because we are carrying out the planning and scheduling ahead of time. Most likely, we are asking for commitment before the resources have been promised elsewhere, because we are making the arrangements with the various matrix managers very early on—with plenty of lead time.

Careful resource allocation during the planning phase prevents schedule slippage later on.

In the course of estimating the time for each activity, we must work with those who will carry out the duties in each case. That is, we must obtain the optimistic, pessimistic, and most likely time estimates from the individuals who will be scheduled to do the work. In the course of estimating these times, it must be known who and how many people of each talent will do the work. Consequently, the resource allocation comes about almost automatically as a result of timing the various activities. After the commitments have been obtained and the various resource managers have documented their commitment by signing-in, then a graphic illustration of this resource allocation is helpful. Sometimes called a *work breakdown structure*, it can be graphed as a histogram, which can be drawn on the same linear time scale as the final draft of the plan network. We strongly advocate doing this.

RESOURCE ALLOCATION FOR THE 50 PERCENT EXPECTED PLAN

We must remember that we needed the rough plan network in order to know how to arrange for and schedule the resources. We must also remember that some will not be available when the original network calls for them; therefore, it will have to be modified in terms of schedule, and perhaps configuration, to suit availability of resources. In the end, the number of people for each activity is based partly on:

1. How many it takes, realistically, to carry out an activity. For example, it is better to have more than one develop a software intensive, electronic hardware product. However, one person is all that is required to inspect the plumbing. In a home under construction, more than one are always required to install the rafters.
2. The schedule desired. People can sometimes be added to the schedule in order to complete an activity sooner. This is not always the case. Too many people are, by definition, a waste of talent.
3. Availability. The number of people desired, based on 2 above, just may not be available, from within the company—all departments considered. It may be desirable to add talent from the outside.

Then, after all managers concur, the network configuration and schedule can be finalized. The reader should refer to Figure 4.4, which illustrates the process that is followed in bringing all the important elements of the *Overall Procedure* into agreement. This process was first presented in Chapter 2. We repeat it here because refining the accuracy and utility of the various major elements of planning and scheduling is extremely important.

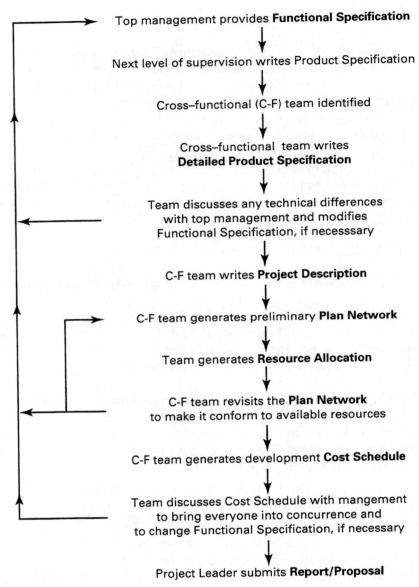

FIGURE 4.4 Flowchart illustration of concurrence.

1. Top management provides the Functional Specification.

2. (a) The next level of supervision, reporting to top management, studies the Functional Specification and writes a Product Specification. This further defines the product and enables identification of a full cross-functional team.

(b) A cross-functional team, which is an expansion of the next level of supervision, of Step 2a above, is formed. This team studies and details, even further, the Product Specification.

At this point in the process, there will probably be some feedback to management. Some facts are known, now, about the product that may influence all the people involved so far to change the Functional Specifications somewhat. For example, in detailing the Product Specification, it may be determined that the development time stated in the Functional Specification was too short.

There will be one or two iterations in this process until the Functional Specifications and the Detailed Product Specification are brought into agreement.

3. The cross-functional team writes the Project Description.

4. The team generates the Plan Network.

5. The team generates the Resource Allocation. It is usually found that resource limitations cause a certain amount of reconfiguring of the Plan Network. Some parallel activities may have to be converted to sequential activities, for example. Therefore, there are one or two iterations back through Step 4 until the network and available resources agree.

Also, since the schedule is now known more accurately, the one stated in the Functional Specification may have to be modified.

6 (a) The cross-functional team generates the development cost estimate. This estimate is based on a hardened schedule of activities and corresponding resources. Therefore, it has great credibility.

(b) The cost stated in the Functional Specification may have to be changed.

Once the above process is fully carried out, we will have all the elements of the *Overall Procedure* in strict agreement, where, when they were first provided, they were not. We can now write the Report.

7. The Project Leader can now write the Report which is a summary of the bottom line results of bringing all the elements of planning and scheduling into concurrence. The Report is actually a request for funding the project. We can, therefore, also think of the Report as a Proposal.

We now turn to the resource allocation chart or histogram. This chart illustrates the work breakdown structure. The features of the chart are the following:

1. There are two sections: the upper section is for the activities on the critical path; the lower section shows the other activities.

2. There is one vertical bar for each activity.

3. The vertical height of each bar is proportional to the number of people allocated to the corresponding activity.

4. The chart is drawn on a linear time scale matching the plan network linear time scale.

5. The horizontal width of each bar is equal to the time expected, t_e, to complete the corresponding activity .

6. For the critical path activities, the left edge of the bar is aligned with the starting event for the corresponding activity, and the right edge is aligned with the activity ending event.

7. For the other activities, the edges of the corresponding allocation bar, in each case, are aligned with the scheduled activity start and finish. Again, the width equals t_e.

8. The event/activity numbers are placed along the time axis in alignment with those corresponding numbers on the plan network.

9. Two other sets of numbers are also located on the resource allocation chart: The calendar scale and the 0, 5, 10—time unit scale. These two scales are identical to the two on the plan network.

10. The name of the specific talent is entered in each bar. For example, "Buyer" is entered in the first bar in Figure 4.8 because "Identify land" is Activity 1–2. LD is entered in the second bar because "Design landscape" is Activity 2–3. In some cases, there is enough space to allow entry of the full activity name. In other cases, an abbreviation must be used. A table adjacent to the chart must explain each abbreviation.

Example 3 from Chapter 3

Work Breakdown Structure for the Printer Replacement Project

For the example of the printer replacement project presented in Chapter 2 and scheduled in Chapter 3, the requirements are the following: two signal processing engineers, one purchasing agent, one hardware engineer, two electronics design engineers, one software engineer, one electronics technician, one systems engineer, one test engineer, and one drafter.

Table 4.1 lists all the various talents required for the printer replacement project.

Figure 4.5 illustrates a section of the human resources for both the critical activities and the non-critical activities.

Figure 4.6 illustrates the complete work breakdown structure for the printer replacement project.

Example 4 from Chapter 3

Work Breakdown Structure for the Home Development Project

For the example of the home building project, the schedule was based on one landscape architect, one surveyor, two excavation people, two drainage plumbers, two foundation workers, and so forth. The complete list of the various talents, and how many of each, is given in Table 4.2.

Refer to Figures 4.7 and 4.8, which have all the information in them, including a legend explaining the abbreviations and two time scales.

The resource allocation chart is a planning and control tool. During the planning cycle, it is used to schedule and gain concurrence on

TABLE 4.1 Number of people for each printer project activity.		
ACTIVITY		NO. OF PEOPLE
1–2	Select printer	2
2–6	Assemble and install printer	1
1–3	Design Input A signal format	1
1–4	Design Input B signal format	1
2–5	Write main printer processing program	1
3–5	Design Input A interface board	1
4–6	Design Input B interface board	1
5–6	Generate databank for output form	1
4–8	Finalize documentation for Channel A	1
5–8	Finalize documentation for Channel B	1
6–8	Test system	1
6–7	Document system and system test	1

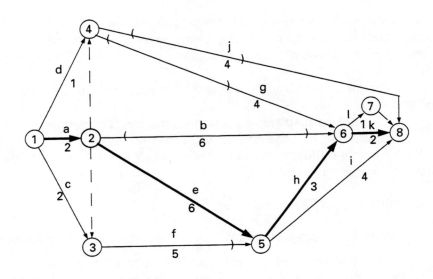

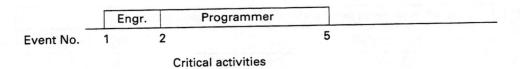

Critical activities

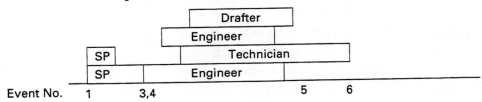

Non-critical activities

FIGURE 4.5 Section of resource allocation chart for critical and non-critical activities for the printer project.

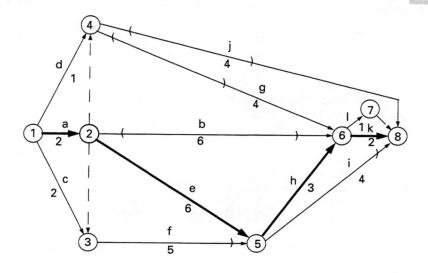

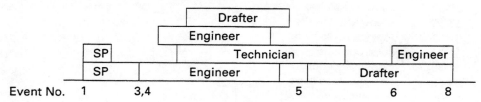

Critical activities

SP=Signal Processing Engineer

Non-critical activities

FIGURE 4.6 Human resources allocation for entire printer replacement project.

TABLE 4.2 Number of people for each activity.

ACTIVITY		NO. OF PEOPLE
1–2	Identify land for home	2
2–3	Landscape design	1
3–4	Survey and locate foundation	1
4–5	Excavate	2
5–6	Underground drainage	2
5–9	Install temporary power	1
6–7	Pour foundation	2
7–8	Insulate basement	2
8–9	Backfill	2
9–10	Plumbing service	2
10–11	Basement plumbing	2
9–13	Frame walls	3
13–14	Install wall sheathing	2
14–15	Install roof trusses	3
11–12	Inside plumbing	2
12–15	Plumbing inspection	1
15–16	Roof sheathing	2
16–17	Shingle roof	3
17–18	Install solar system	2
9–19	Fireplace	1
19–20	Chimney and chimney flashing	1
18–20	Pour basement floor	3
20–21	Install HVAC system	2
20–24	Windows and exterior doors	2
24–28	Install siding	2
21–22	Wiring	1
22–23	Security system	1
23–29	Wiring inspection	1
20–25	Grade lot	2
25–26	Driveway and sidewalk	2
26–27	Plant shrubs	3
27–28	Lay sod	3
29–30	Insulate walls and ceilings	2
30–31	Sheet rock	2
31–32	Taping and skim coat	2
32–34	Install cabinets	2
32–33	Install interior doors	2
34–35	Lay hardwood floors	1
35–36	Sand floors	1
36–37	Finish floors	1
37–38	Interior finishing	2
38–39	Light fixtures and plumbing fixtures	2
28–39	Exterior finish	3
39–40	Clean windows	1
40–41	Clean property	1

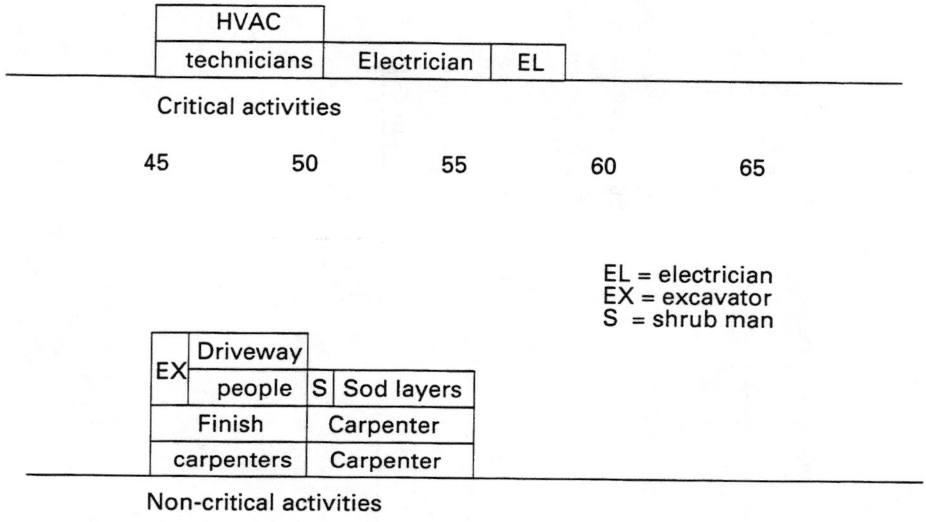

FIGURE 4.7 Section of complete resource chart for home development project.

resource allocation, and for budgeting. After project approval, and during the whole time when the project is being carried out, the chart is used for control. The project leader, cross-functional team and matrix resource managers can all see, at a glance, when each and every one of the talents will be needed. They can also see far enough ahead to make sure everybody will be available when needed. At each review meeting, the team will place the plan network and the resource allocation chart side by side, aligned in time, and their whole project can be seen before them. This process is excellent for concurrent planning, scheduling, and controlling. It is the essence of concurrent project management.

RESOURCES OTHER THAN PEOPLE

So far we have dealt with the most important resources: the people who actually carry out the activities. Many other resources are required besides people in most projects: test equipment, prototyping materials, office space, laboratory space, engineering workstations and other capital equipment, books for research, and clerical support.

Each project has its own special requirements and they must all be taken into account, planned on and scheduled. Suppliers of these resources must all concur. See Tables 4.3 and 4.4, which list all the various materials needed.

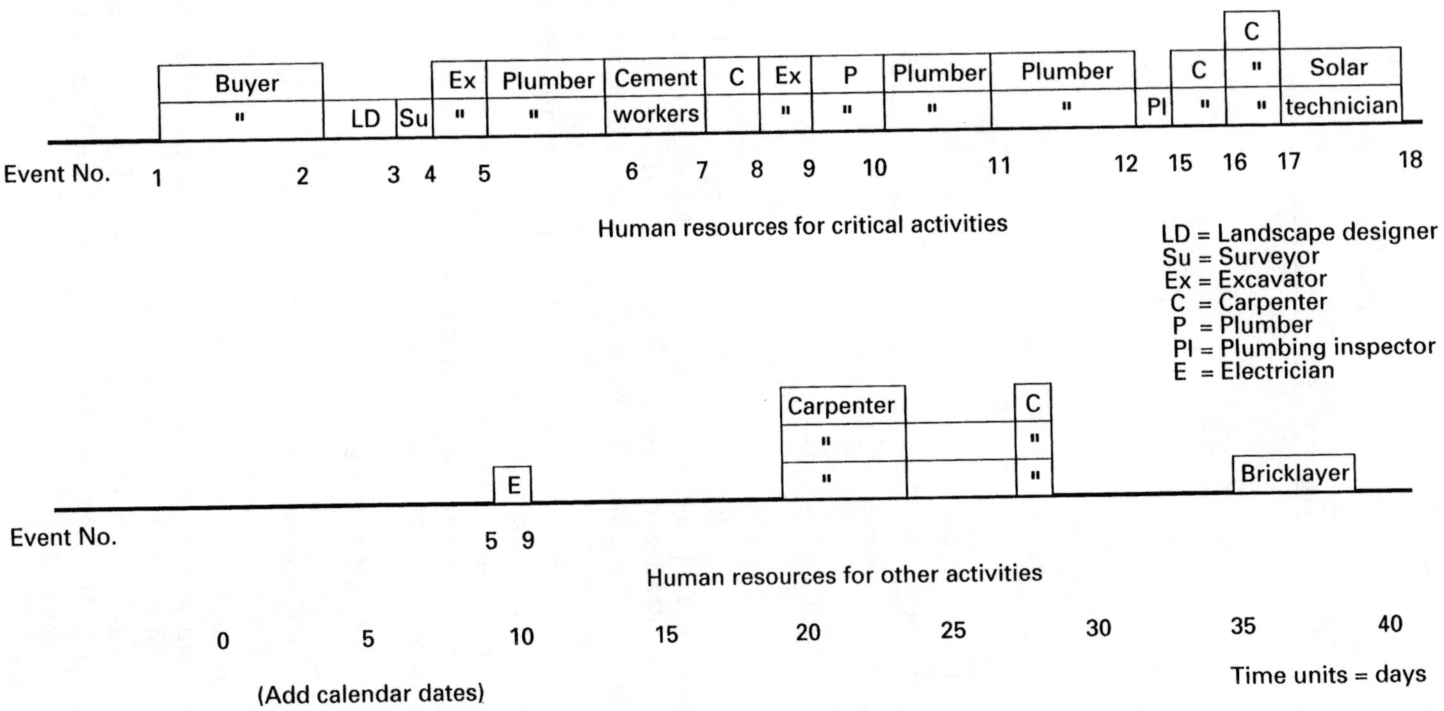

FIGURE 4.8A Human resource allocation for development of home.

144

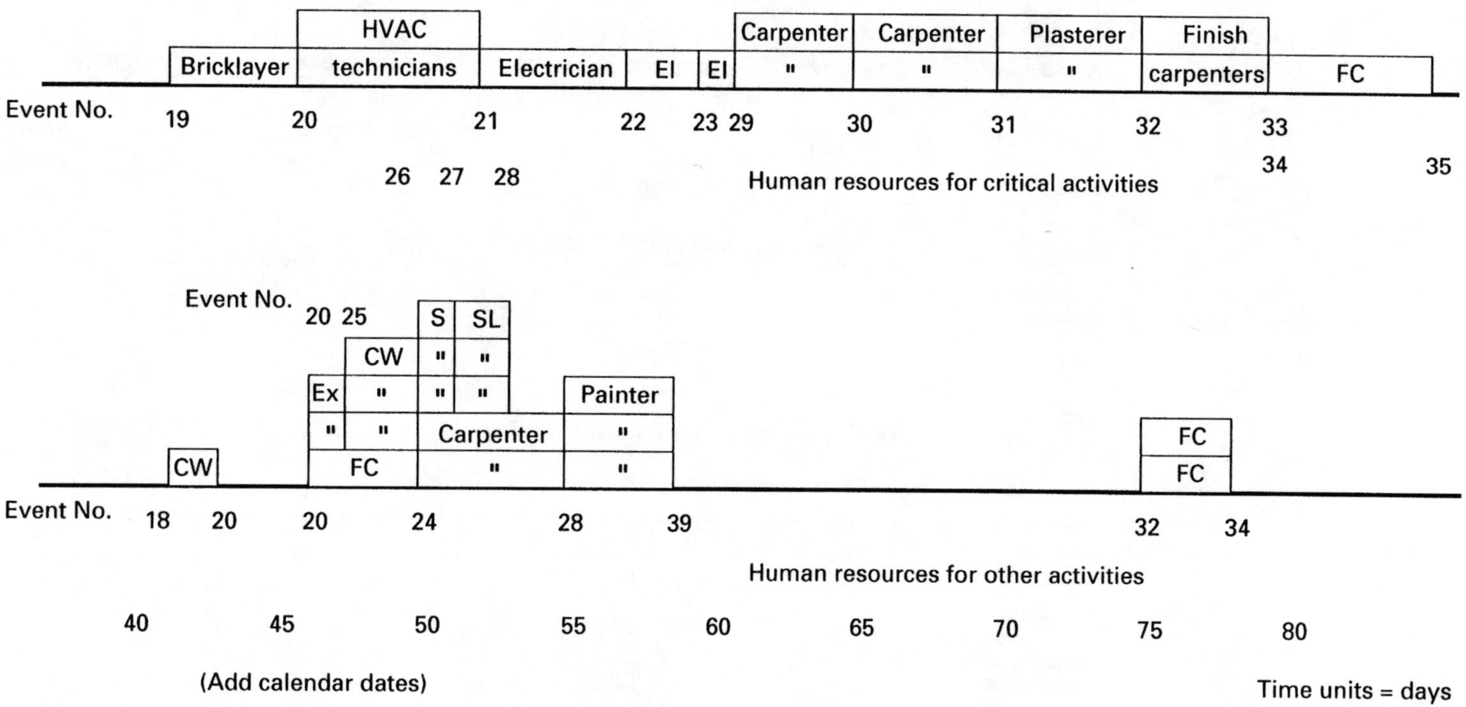

FIGURE 4.8B Human resource allocation for development of home.

Time units = days

Human resources for critical activities

Human resources for other activities

(Add calendar dates)

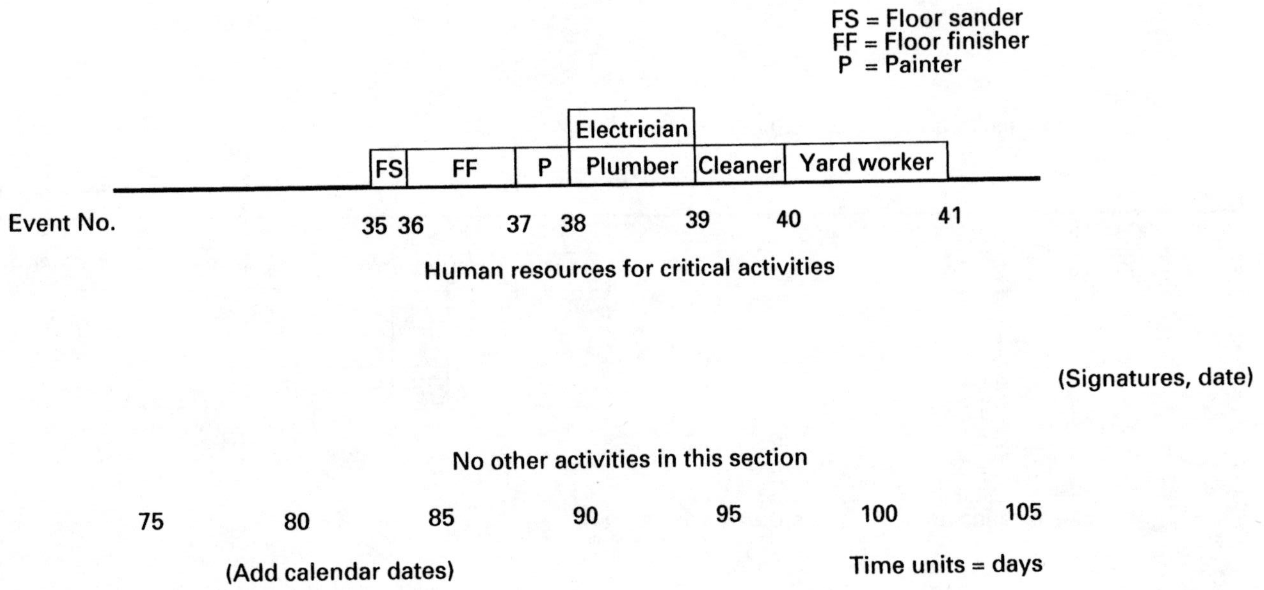

FS = Floor sander
FF = Floor finisher
P = Painter

(Signatures, date)

Event No.

| FS | FF | P | Plumber | Cleaner | Yard worker |

Electrician

| 35 | 36 | 37 | 38 | 39 | 40 | 41 |

Human resources for critical activities

No other activities in this section

(Add calendar dates)

Time units = days

| 75 | 80 | 85 | 90 | 95 | 100 | 105 |

FIGURE 4.8C Human resources for development of home.

TABLE 4.3 Materials required for printer project.

OTHER RESOURCES REQUIRED FOR PRINTER REPLACEMENT PROJECT

(a) specifications
information regarding available printers
Thomas Register
office
utilities
transportation/meals/lodging
space for testing printers
system in which to integrate printer for testing
printer paper and ribbons
instruction manual
test protocols
test equipment

(b) tools
lab space
utilities
manuals
system in which to integrate printer
interface/adaptation components
printer
printer paper/ribbons
safety equipment

(c,d) office
lab space
utilities
engineering workstation
software
specifications
test equipment
calculator
engineering notebook

(e) office
utilities
computer
software

TABLE 4.3 Materials required for printer project (Cont.)

OTHER RESOURCES REQUIRED FOR PRINTER REPLACEMENT PROJECT

specifications for selected printer
system specifications
manuals
emulation board
lab space
rented printer
test equipment
documentation notebook

(f,g) office
lab space
utilities
engineering workstation
software
specifications
manuals
input from signal format designers
engineering notebook
prototype components
tools, solder, wire-wrapping hardware
test equipment

(h) office
lab space
utilities
specifications
input from board designer
computer
engineering notebook

(i) completely installed system
lab space
utilities
test equipment
test procedure
manuals
test equipment
tools
documentation notebook

TABLE 4.4 Materials required for the home development project.

ACTIVITY		NO. OF PEOPLE	MATERIAL
1–2	Identify land for home	2	
2–3	Landscape design	1	Art materials
3–4	Survey and locate foundation	1	Transit
4–5	Excavate	2	Bulldozer
5–6	Underground drainage	2	Pipes and shovels
5–9	Install temporary power	1	Switch and wire
6–7	Pour foundation	2	Cement and tools
7–8	Insulate basement	2	Insulation
8–9	Backfill	2	Bulldozer
9–10	Plumbing service	2	Plumbing supplies
10–11	Basement plumbing	2	Plumbing supplies
9–13	Frame walls	3	Lumber, nails, tools
13–14	Install wall sheathing	2	Sheathing, nails, tools
14–15	Install roof trusses	3	Trusses, nails, tools
11–12	Inside plumbing	2	Plumbing supplies
12–15	Plumbing inspection	1	
15–16	Roof sheathing	2	sheathing, nails, tools
16–17	Shingle roof	3	shingles, nails, tools
17–18	Install solar system	2	Solar system, tools
9–19	Fireplace	1	Bricks, tools
19–20	Chimney and chimney flashing	1	Bricks, flashing, tools
18–20	Pour basement floor	3	Cement, tools
20–21	Install HVAC system	2	HVAC system, tools
20–24	Windows and exterior doors	2	Windows, doors, tools
24–28	Install siding	2	Siding, tools
21–22	Wiring	1	Wire, boxes, tools
22–23	Security system	1	Security system, tools
23–29	Wiring Inspection	1	
20–25	Grade lot	2	Bulldozer, rakes
25–26	Driveway and sidewalk	2	Cement, tools
26–27	Plant shrubs	3	Shrubs, tools
27–28	Lay sod	3	Sod, tools
29–30	Insulate walls and ceilings	2	Insulation
30–31	Sheet rock	2	Sheet rock, tools
31–32	Taping and skim coat	2	Tape, plaster, tools
32–34	Install cabinets	2	Cabinets, tools
32–33	Install interior doors	2	Doors, tools
34–35	Lay hardwood floors	1	Flooring, tools

TABLE 4.4 Materials required for the home development project. (Cont.)

ACTIVITY		NO. OF PEOPLE	MATERIAL
35–36	Sand floors	1	Sandpaper, machine
36–37	Finish floors	I	Polyurethane, brushes
37–38	Interior finishing	2	Paint, solvent, brushes ladders, masking tape
38–39	Light fixtures and plumbing fixtures	2	Fixtures, tools
28–39	Exterior finish	3	Paint, brushes
39–40	Clean windows	1	Cleaner, clothes
40–41	Clean property	1	Rakes, bags

The resources other than personnel are not as dependent upon approval for availability at prescribed times as the personnel resources are. Regarding personnel, it may be relatively difficult to obtain guaranteed approval for a certain person to be available to carry out a specific activity on October 20, for example, for two weeks. It is tough to do this six months prior to October 20, when the project is being scheduled. Yes, it is necessary to obtain signed approval for this to happen, but it is a troublesome task, relative to scheduling other resources. Although lead times for the other resources must be known to determine realistic scheduled times, it is easy compared with scheduling people.

At any rate of difficulty, it all must be done. Without addressing these issues, projects are not scheduled very carefully or realistically. These nonrealistic schedules are impossible to control, so we must carry out the scheduling of not only the people but the other resources.

When we have the complete set of resources scheduled, then we can draw a picture of them as an allocation chart.

1. Do this for the critical activities.
   (a) for the people
   (b) for the other resources
2. Do this for the remaining activities.
   (a) for the people
   (b) for the other resources

See Figure 4.9.

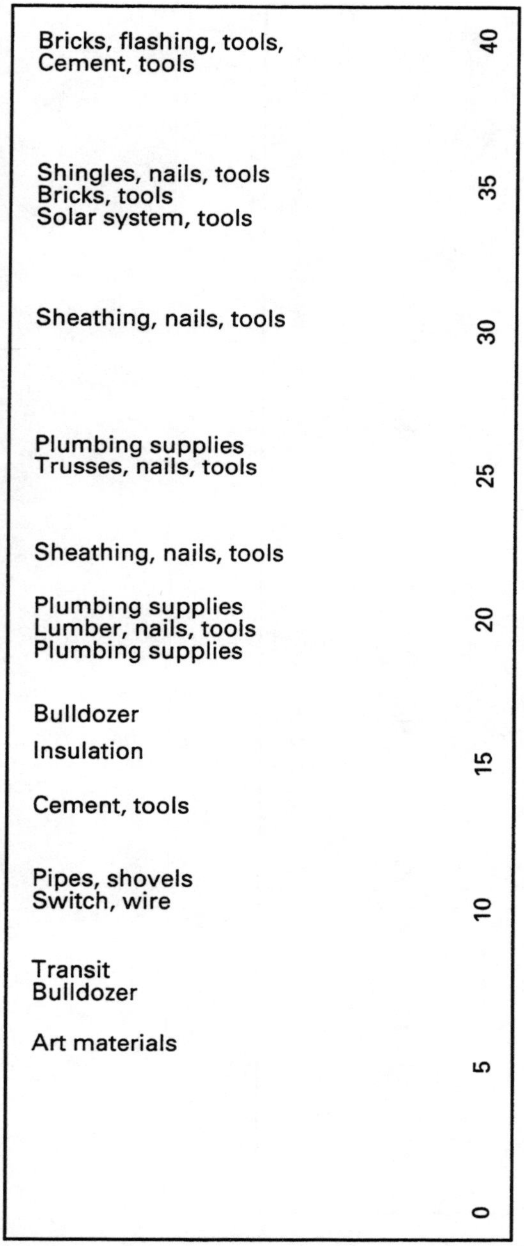

Bricks, flashing, tools,
Cement, tools

40

Shingles, nails, tools
Bricks, tools
Solar system, tools

35

Sheathing, nails, tools

30

Plumbing supplies
Trusses, nails, tools

25

Sheathing, nails, tools

Plumbing supplies
Lumber, nails, tools
Plumbing supplies

20

Bulldozer

Insulation

15

Cement, tools

Pipes, shovels
Switch, wire

10

Transit
Bulldozer

Art materials

5

0

Flooring, tools

80

Cabinets, tools
Doors, tools

75

Tape, plaster, tools

70

Sheet rock, tools

65

Insulation

60

Security system, tools

55

Siding, tools
Shrubs, tools
Sod, tools
Wire, boxes, tools

50

HVAC system, tools
Windows, doors, tools
Bulldozer, rakes
Cement, tools

45

40

FIGURE 4.9A Resource allocation for non-personnel resources.

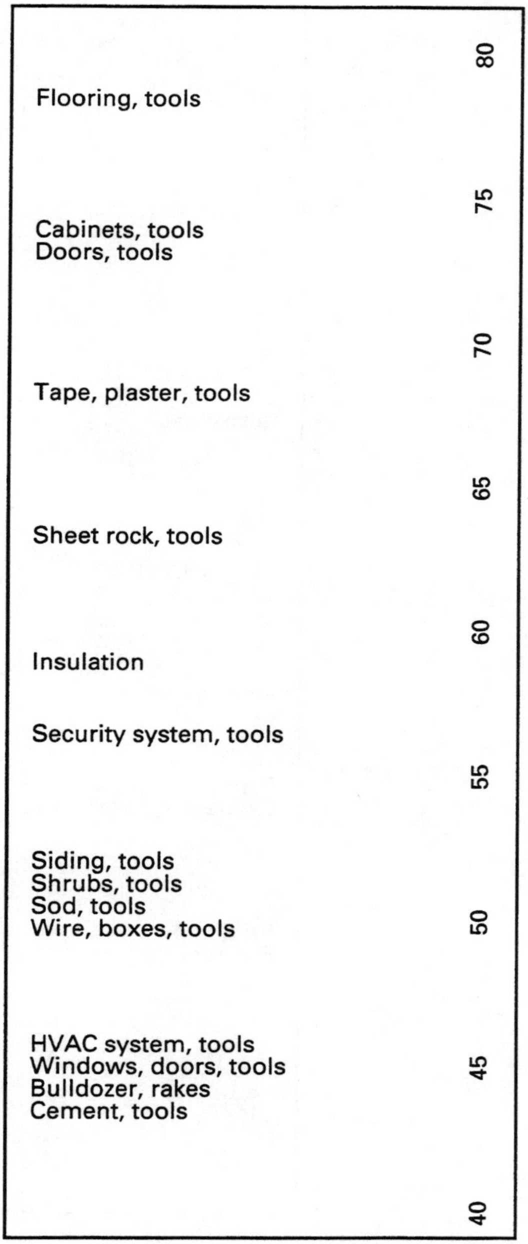

FIGURE 4.9B Resource allocation for non-personnel resources.

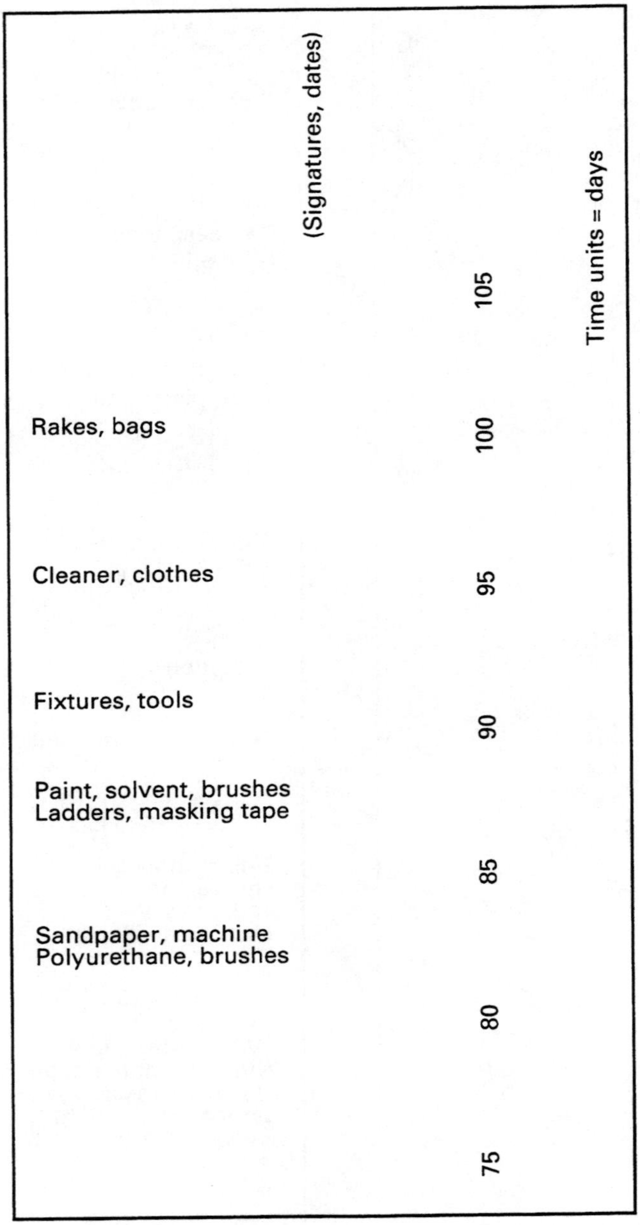

(Signatures, dates)

Time units = days

Rakes, bags

105

100

Cleaner, clothes

95

Fixtures, tools

90

Paint, solvent, brushes
Ladders, masking tape

85

Sandpaper, machine
Polyurethane, brushes

80

75

FIGURE 4.9C Resource allocation for non-personnel resources.

It can be argued that there are many other steps, substeps, subactivities, and tasks that can be added to our plan. However, I feel that what we have here as a procedure and for illustrations is complex enough.

I find in dialoguing with people from industry that, in most cases, not many people in their respective companies even do this much. So why present an overall procedure and generate charts, graphs, and other illustrations that are more complicated than these? When people follow this relatively simple procedure, that, in itself, is a major step forward. Industry people involved in projects consider this a major improvement in project management. Attempting to implement a more complex procedure decreases the probability that there would be any change for the better at all. They would feel overwhelmed at the outset, and would be prone to sticking to their old, unstructured habits of forging ahead without sufficient planning.

We have found it intriguing and challenging enough to merely take the level of project management within their respective companies from where they were to the level presented in this book.

A resource allocation process that is too complex is self-defeating.

Additional complexity would be self-defeating because it would be overwhelming from the outset and therefore would not be productive in improving the concurrent project management process within the companies. In other words, keep the process as simple as possible and yet make a major improvement in concurrent project management.

Next, then, let us further harden the plan network and resource allocation. They have now been brought into agreement with all the resources on the list, each and every one guaranteed for availability at the required times and signed for by all the members of the cross-functional team. One other major step must be taken. So far we have been working on the expected (50 percent) plan network. We know the 95 percent scheduled time must be addressed.

RESOURCE ALLOCATION FOR THE 95 PERCENT PROBABLE PLAN

When we "stretch" the plan network to reflect the 95 percent situation, the resources will be required at slightly different times in some cases. Availability must be checked for these 95 percent times. See Figure 4.10. This is a segment of our home development project.

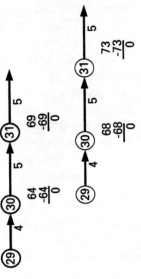

FIGURE 4.10 Comparison of 50 percent schedule with 95 percent schedule.

Now, the question is: Will the resources required for Activity 30–31 be available for the new activity period on time? Chances are, they will be because the resource is being planned with a longer lead time, by definition. The 95 percent times are always further away than the 50 percent times; therefore, the lead times are longer. But the availability must be checked anyway, because the times are, in fact, somewhat different than those for which the previous commitments were made. Once they are checked and approved, we again have the whole team in concurrence.

SUMMARY

We have now hardened a plan, yet it is still only a plan, a projection of what will happen in the future. However, it is a hardened plan and one that is well structured, based on consensus and concurrence by all who have been and will be involved.

This plan includes a plan network that is a very graphic illustration of the activities that compose the project, clearly showing those in sequence and others in parallel. It includes an allocation of all the resources required to carry out the project, each of which is shown in graphic form in alignment with the corresponding activity, in each case. For example, the resources for Activity 37–38, interior finishing, are two painters, step ladders, brushes, paint, solvent, and masking tape. These resources are illustrated in Figure 4.9, and when these two figures are aligned together so that the resource allocation is directly below the plan network with the time scales aligned, it can be seen that the resources for Activity 37–38 are directly below Activity 37–38 in the network.

It may seem trivial that masking tape is listed. But this is merely an example, one which is presented to make a point. The parallel in our next project may be integrated circuits required for prototyping. There is often a very long lead time required for some circuits. The point is that *all* resources must be listed and checked, for some will present problems later if not properly addressed during the planning stages.

We now have completed Step 5 in the *Overall Procedure*.

The *Overall Procedure* is shown, again, on the next pages with the completed steps in **bold**. The next major step is that of computing the projected cost to be incurred by the company in carrying out the project.

This will be done at least two different ways: first for the expected (50 percent) schedule and, second, for the 95 percent probability schedule. If desired, it can be done for any other probability between 50 percent and 100 percent.

OVERALL PROCEDURE

1. Clarify the *FUNCTIONAL SPECIFICATIONS*. Study the elements of planning to make sure you and your teammates have properly addressed all aspects of planning. Clarify the company objectives and the marketing objectives of the new product development. Study competition. Determine customers' needs. Write product description. Write final functional specifications. Document agreement with management team by everyone on the team signing the final specifications.

2. Detail the product specifications. Identify and establish the cross-functional team. Study competition. Study the functional specification. Select the design concept. Detail all aspects of the physical product. Rely on experts. Write a *DETAILED PRODUCT DEFINITION*. Document agreement within the cross-functional team by having all members of the team sign the detailed product definition.

3. Describe the project. Elements 1 and 2, above, describe the product. This element, the *PROJECT DESCRIPTION*, describes the project of developing the product. List all the tasks composing the project—list approximately forty or fifty tasks. Rely on experts. Write a detailed project description. Document agreement within the team in the same manner as above.

4. Generate the *PLAN NETWORK*. List all the tasks/activities that compose your project. Prioritize them. Indicate after each activity the one or more other activities that immediately precede the one under study. Draw a rough network to illustrate the correct configuration. Rely on experts. Study resources. All members of the cross-functional team must sign the plan network.

5. Validate the *RESOURCE ALLOCATION*. Draw the work breakdown structure and assign the different talents to corresponding tasks. Obtain approval from the source managers for scheduled allocations. Modify the plan network configuration if necessary. *Redraw the plan network, on a linear time scale.* Identify the critical path. All members of the cross-functional team must sign the revised plan network.

   Draw the resource allocation in the form of a histogram/vertical bar chart on the same time scale as used for the plan network. All members of the cross-functional team must sign the resource chart.

6. Generate the **COST SCHEDULE**. Draw a cost curve as a function of time using the same time scale as used for the plan network and the resource histogram. All members of the cross-functional team must sign the cost curve.

7. **REPORT** the plan to management and request funding. Because experts were relied upon for the above planning, the best possible plan has been put in place. The project leader has the responsibility of presenting the plan for approval.

Figures 4.11 and 4.12 illustrate the final result of our time and resource scheduling. They embody:

- a graphical illustration of our project showing each activity in relationship to all the other activities
  - activity priority
  - activity completion date
- the critical path

FIGURE 4.11 Printer project network with personnel assignments.

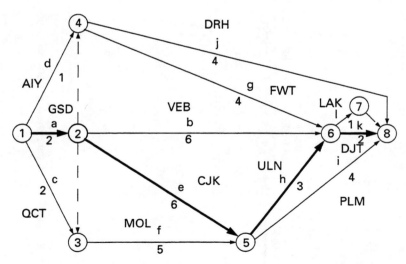

Must add event times, linear time scale, calendar scale, and critical path

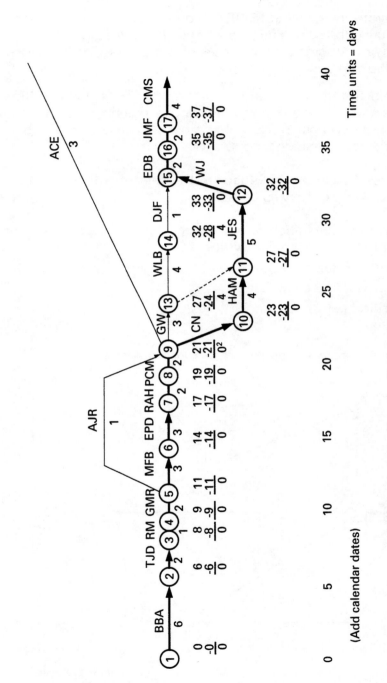

FIGURE 4.12A Plan Network for development of home with assignments.

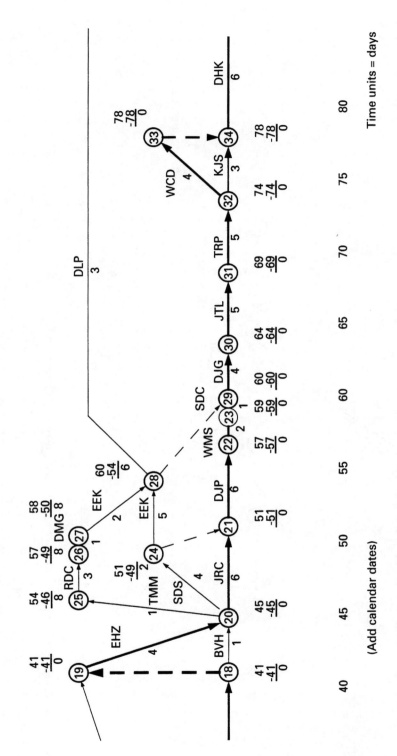

FIGURE 4.12B Plan Network for development of home with assignments.

159

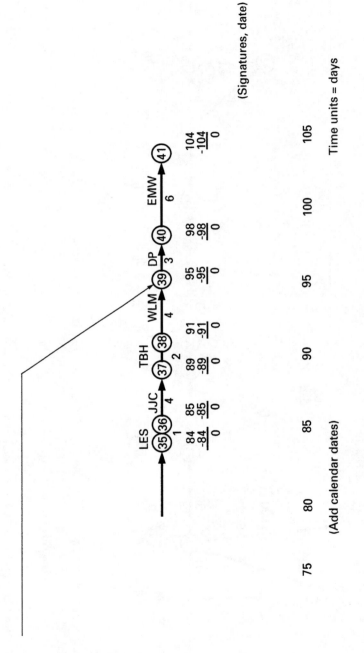

FIGURE 4.12C Plan Network for development of home with assignments.

- who will carry out each activity
  - implication of commitment
- overall time scale
- calendar dates

PROBLEMS

4.1 It has been found in the project of buying an automobile that the plan network is that of Figure 2.8, because of limited resources. Draw a resource allocation chart for this case. Use the activity times given in Problem 2.10.

4.2 Draw a second resource allocation chart matching Figure 2.7 by using the same activity times as in Problem 2.10.

4.3 Why is it necessary to obtain approval from the heads of the departments from which human resources will be drawn in the future, in the case of matrix resourcing?

4.4 In addition to human resources, list several other resources one would need when constructing and testing a prototype electromechanical product.

4.5 Why might it be necessary to modify the plan network after the resource study is made.

4.6 Given the following rough draft network, construct the matching resource allocation chart. Be sure to include all elements of the chart as stipulated in Chapter 4. Put the chart on a linear time scale.

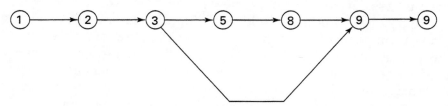

Plan network for buying an automobile.

4.7 Assume that rescheduling the project in Problem 4.6 for a 95 percent probability results in shifting all activity completion events out by 20 percent. Draw the corresponding resource allocation chart.

5 COST

INTRODUCTION

The actual cost of a project is the total cost incurred in actually carrying it out. If we are discussing a new product development, then the total cost is the research and development, product engineering, quality engineering, manufacturing engineering, marketing development, and the procurement of capital equipment, space, and all the remaining elements of the business development of the new product venture.

Every new product should be validated with an approved business plan. A sample is given on page 163. Consideration of all the elements of this plan is beyond the scope of this book. Suffice it to say that every new product development project should be based on a sound business plan. At times, a financial expert from within the company ought to be part of the team to infuse adequate business discipline into the project.

In this book we will limit our discussion to the cost of the activities carried out in the technical development of the product, not the marketing development. The example we use is the development of the home.

I am presenting the planning, scheduling, and controlling of a project in this book. It may be that of developing a new product, and we often make reference to that case herein. It may be the upgrading of an old product. In

Company _____
Project Name: _____
Case/Plan # _____
Start Date _____
Rev. Date _____

Date: _____
Total Capital _____

Submission _____
Status Report _____
Product Plan _____
Budget Planning _____

Category
Major Prod _____
Minor Prod _____

Corporate approval:
Product Dev. _____
Executive _____

Unit Approval:
Finance _____
Engineering _____
Marketing _____
Manuf. _____
Gen Mgmt. _____

Years	Yr	Yr	Yr	Yr	Yr	Yr	Yr	Total
Total Market Unit Sales		100	625	700	700	650	540	3315
Sales		100	625	700	700	650	540	3315
STD Costs		53	330	370	370	345	286	1754
Incremental Sales		33	208	233	233	217	180	1104
Incremental STD Cost		17	104	117	117	109	90	554
Change in STD Costs								
Contributed Margin		63	399	446	446	413	344	2111
Development Eng	162	130						292
Support Eng								
Marketing		80	60	40	40	40	40	300
G & A		5	33	37	37	35	29	176
Depreciation		5	8	8	7	7		35
Other								
Contributed Income	-162	-157	298	361	362	331	275	1308
FIT	-76	-72	137	166	166	152	127	600
Net Contr. Income	-86	-85	161	195	196	179	148	708
Depreciation		5	8	8	7	7		35
Capital Equipment		35						35
Incremental WC		34	182	15	0	-16	-39	176
Other								
Net Cash	-86	-149	-13	188	203	202	187	532
ROI								
Net Income % of Sales								
Cash	-86	-235	-248	-60	143	345	532	

163

general, we are discussing any technical project, although most of the elements and procedures discussed cover nontechnical ones as well. However, let us focus on concurrent project management of technical projects.

GENERATION OF THE PREDICTED COST

We will generate a cost curve under the following rules:

1. List out the cost per unit of each of the various talents required. Consider only the labor cost, for now. Material, equipment, and space costs will be added later.

2. Identify the points in time at which certain important activities are scheduled to be completed. Label these as milestones. The objective here is to compute the project cost scheduled to be incurred at each of the milestones. For example, important milestones for the home development project might be: completion of foundation and basement, completion of framing, completion of plumbing and wiring, or, in an industrial hardware, software product, the completion of printed circuit design, completion of housing design, completion of software, completion of documentation, completion of testing, and so forth.

3. Compute the accumulated expected cost up to each milestone.

4. Add the cost of all the activities, including critical path activities, and any other activities that were scheduled for completion before the milestone in each case.

5. Plot the expected cost at each milestone on a graph with the same time scale that was used for the final plan network and the resource allocation chart.

6. Add the costs of the resources other than people, such as material, equipment, and space. The sum of these and the labor can be plotted as a total cost curve.

7. Add the calendar scale and the time unit scale—as on the other two documents.

The cost of personnel resources required for all the various activities will be computed first. This is done by starting with a copy of the list of activities, such as in Table 4.1 or 4.2.

Adjacent to each activity, and the corresponding number of required people with that specific talent, enter the cost per day that will be incurred as an additional project cost by having that person or those persons working during the activity period. Do this for each and every activity. The sum of all those personnel costs will be the total. See Table 5.1.

Then plot a curve illustrating the cost schedule.

It is recommended that enough points be plotted so that everyone on the team can easily see the cost schedule.

The cost curve should be plotted on the same linear time scale that the plan network and the resource allocation are plotted on.

So far we have been addressing personnel costs only. We might wish to compute the total cost in steps also. We will figure the cost for the personnel resources first and plot it. Then we will plot the cost for the other resources. Third, we will add them together and obtain a curve or schedule for the total cost.

In Figure 5.1, the personnel costs are shown as the thin solid line; the other costs are shown as the dashed line, and the total as the thick solid line. Later, in Chapter 8, we will add the actual cost as the thick-dashed curve on the same chart. This will provide the team with a pictorial illustration/comparison between the scheduled cost and the actual cost as a primary control tool.

The complete list of activities with costs other than personnel is given in Table 5.2.

The next question is how many points should be plotted and which ones. We can plot a point every time the plan network indicates a major milestone, or a point matching the end of every week or month in the schedule. Or we can plot a point for every $10,000 or $25,000 of expected cost. The object is that we want to schedule cost at frequent enough intervals so that we have a measure of our performance later when the project is under way and we are controlling it.

In our example of the development of the home, we have chosen six major milestones. These are the completions of Activities 6–7, 12–15, 18–20, 27–28, 32–33, and 40–41. The costs for each of these are totaled. The subtotals are also given in Table 5.1. Each of these is a point to be plotted, and are shown as the thin solid line curve in Figure 5.1.

The costs for the other resources are listed in Table 5.2, and the subtotals are given for each of these as the same milestones used for Table 5.1. These are plotted as the dashed line in Figure 5.1.

The total cost, the sum of the thin solid line curve and the dotted line curve, is plotted as the heavy solid line curve in Figure 5.1.

TABLE 5.1 Activities in normal order with expected times and cost.

ACTIVITY		NO. PEOPLE	t_e	$/DAY/ PERSON	$ ACTIVITY	TOTAL COST IN $
1–2	Identify land for home	2	6	---		
2–3	Landscape design	1	2	300	600	600
3–4	Survey and locate foundation	1	1	200	200	800
4–5	Excavate	2	2	120	480	1280
5–6	Underground drainage	2	3	200	1200	2480
5–9	Install temporary power	1	1	200	200	2680
6–7	Pour foundation and basement walls	2	3	120	720	**3400 = Cost$_1$**
7–8	Insulate basement	2	2	150	600	4000
8–9	Backfill	1	2	120	240	4240
9–10	Plumbing service	2	2	200	800	5040
10–11	Basement plumbing	2	4	200	1600	6640
9–13	Frame walls	3	3	150	1350	7990
13–14	Install wall sheathing	2	4	150	1200	9190
14–15	Install roof trusses	3	1	150	450	9640
11–12	Inside plumbing	2	5	200	2000	11640
12–15	Plumbing inspection	1	1	—		**11640 = Cost$_2$**
15–16	Roof sheathing	2	2	120	480	12120
16–17	Shingle roof	3	2	120	720	12840
17–18	Install solar system	2	4	200	1600	14440
9–19	Fireplace	1	3	200	600	15040
19–20	Chimney and chimney flashing	1	4	200	800	15840
18–20	Pour basement floor	3	1	120	360	**16200 = Cost$_3$**
20–21	Install HVAC system	2	6	200	2400	18600
20–24	Windows and exterior doors	2	4	200	1600	20200
24–28	Install siding	2	5	150	1500	21700
21–22	Wiring	1	6	200	1200	22900
22–23	Security system	1	2	200	400	23300
23–29	Wiring Inspection	1	1	—		
20–25	Grade lot	2	1	200	400	23700
25–26	Driveway and sidewalk	2	3	120	720	24420
26–27	Plant shrubs	3	1	120	360	24780
27–28	Lay sod	3	2	120	720	**25500 = Cost$_4$**
29–30	Insulate walls and ceilings	2	4	150	1200	26700
30–31	Sheet rock	2	5	150	1500	28200
31–32	Taping and skim coat	2	5	200	2000	30200
32–34	Install cabinets	2	3	200	1200	31400
32–33	Install interior doors	2	4	200	1600	**33000 = Cost$_5$**
34–35	Lay hardwood floors	1	6	200	1200	34200
35–36	Sand floors	1	1	150	150	34350
36–37	Finish floors	1	4	150	600	34950
37–38	Interior finishing	2	2	150	600	35550
38–39	Light fixtures and plumbing fixtures	2	4	200	1600	37150
28–39	Exterior finish	2	3	150	900	38050
39–40	Clean windows	1	3	120	360	38410
40–41	Clean property	1	6	120	720	**39130 = Cost$_6$**

TABLE 5.2 Cost of materials required.

ACTIVITY		NO. OF PEOPLE	MATERIAL	COST	TOTAL COST IN $
1–2	Identify land for home	2			
2–3	Landscape design	1	Art materials	250	250
3–4	Survey and locate foundation	1	Transit	150	400
4–5	Excavate	2	Bulldozer	800	1200
5–6	Underground drainage	2	Pipes and shovels	1000	2200
5–9	Install temporary power	1	Switch and wire	250	2450
6–7	Pour foundation	2	Cement and tools	1250	**3700**
7–8	Insulate basement	2	Insulation	750	4450
8–9	Backfill	2	Bulldozer	400	4850
9–10	Plumbing service	2	Plumbing supplies	800	5650
10–11	Basement plumbing	2	Plumbing supplies	2300	7950
9–13	Frame walls	3	Lumber, nails, tools	12500	20450
13–14	Install wall sheathing	2	Sheathing, nails, tools	2500	22950
14–15	Install roof trusses	3	Trusses, nails, tools	7500	30450
11–12	Inside plumbing	2	Plumbing supplies	3300	33750
12–15	Plumbing inspection	1			**33750**
15–16	Roof sheathing	2	sheathing, nails, tools	1500	35250
16–17	Shingle roof	3	shingles, nails, tools	1750	37000
17–18	Install solar system	2	Solar system, tools	4750	41750
9–19	Fireplace	1	Bricks, tools	1250	43000
19–20	Chimney and chimney flashing	1	Bricks, flashing, tools	1250	44250
18–20	Pour basement floor	3	Cement, tools	1600	**45850**
20–21	Install HVAC system	2	HVAC system, tools	7500	53350
20–24	Windows and exterior doors	2	Windows, doors, tools	1500	54850
24–28	Install siding	2	Siding, tools	2000	56850
21–22	Wiring	1	Wire, boxes, tools	3000	59850
22–23	Security system	1	Security system, tools	750	60600
23–29	Wiring Inspection	1			
20–25	Grade lot	2	Bulldozer, rakes	1600	62200
25–26	Driveway and sidewalk	2	Cement, tools	4200	66400
26–27	Plant shrubs	3	Shrubs, tools	2200	68600
27–28	Lay sod	3	Sod, tools	1700	**70300**
29–30	Insulate walls and ceilings	2	Insulation	750	71050
30–31	Sheet rock	2	Sheet rock, tools	900	71950
31–32	Taping and skim coat	2	Tape, plaster, tools	1800	73750
32–34	Install cabinets	2	Cabinets, tools	1200	74950
32–33	Install interior doors	2	Doors, tools	950	**75900**
34–35	Lay hardwood floors	1	Flooring, tools	1900	77800
35–36	Sand floors	1	Sandpaper, machine	500	78300
36–37	Finish floors	1	Polyurethane, brushes	600	78900
37–38	Interior finishing	2	Paint, solvent, brushes ladders, masking tape	1200	80100
38–39	Light fixtures and plumbing fixtures	2	Fixtures, tools	3700	83800
28–39	Exterior finish	3	Paint, brushes	1500	85300
39–40	Clean windows	1	Cleaner, clothes	100	85400
40–41	Clean property	1	Rakes, bags	50	**85450**

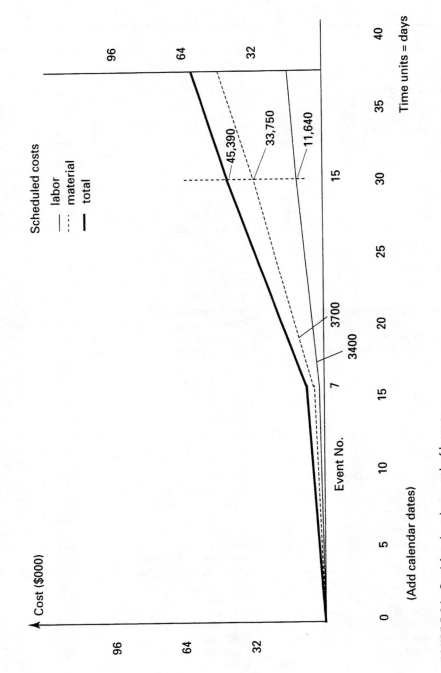

FIGURE 5.1A Cost for development of home.

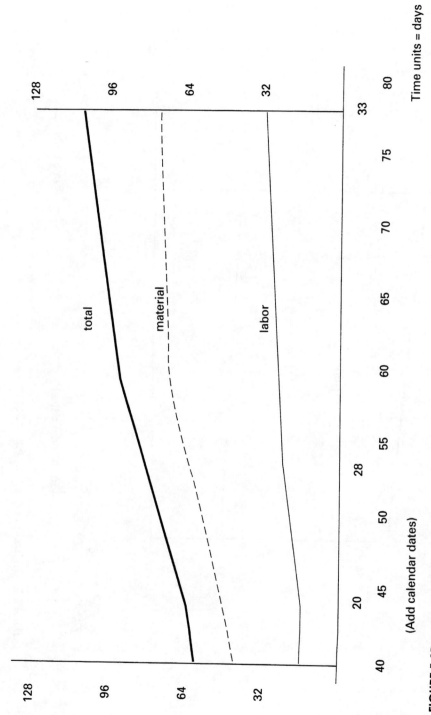

FIGURE 5.1B Cost for development of home.

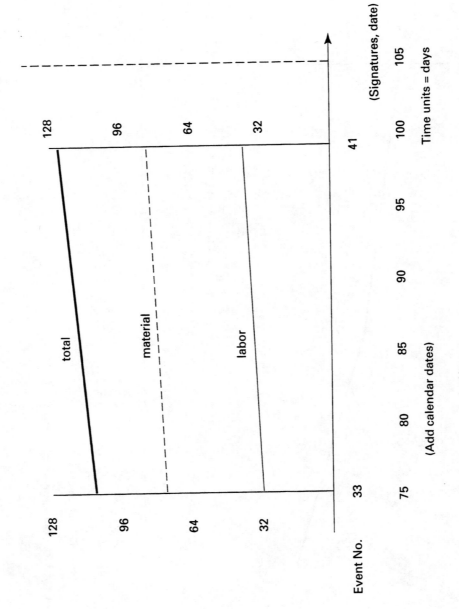

FIGURE 5.1C Cost for development of home.

We have now completed the cost schedule defined as the expected cost. This is true because we have used the expected activity times in generating the cost schedule. Again, we must realize as a team, and we must report, that this is the cost we expect to complete the project with. However, we must also report the very important fact that there is a 50 percent chance of spending more than this computed cost.

The next section will cover the calculation of the 95 percent probable cost. Reference is made to the section in Chapter 3 dealing with the more meaningful project scale.

THE MORE MEANINGFUL COST ESTIMATE

So far the cost has been computed for each activity by using the expected time, t_e, for that activity. This process resulted in the *expected* cost for each. We know from Chapter 3 that there is a 50 percent chance of exceeding this expected cost.

Also, in Chapter 3, we calculated the project schedule corresponding to a 95 percent probability of success. Similarly, we will calculate a project cost corresponding to a 95 percent probability of success. Basically, the 95 percent cost is derived from the 95 percent plan network and from the 95 percent resource allocation discussed in Chapters 3 and 4.

There are various ways of calculating the 95 percent project cost.

1. *Proportional*—simple proportions.

   The simplest calculation, and one which is just a first approximation, is based on the simple ratio of the 95 percent time schedule to the critical path schedule. That is, the 95 percent probable cost is

   $Cost = (ST_{95}/ST_{50}) \times$ expected cost

   In the case of the home building project, this works out to be

   $Cost_{95} = (110/104) \times \$39,130 = \$41,388$

   An equivalent way of viewing this procedure is by thinking of the percentage increase.

   Therefore, if the 95 percent time schedule is 110 days, then it can be shown that this is 5.77 percent longer than the expected time of 104 days.

To find the 95 percent cost, add 5.77 percent to the expected cost. Therefore, the 95 percent cost would be

$39,130 + 5.77 percent = $41,388

We are dealing with personnel costs, only, here. A parallel procedure can be carried out for the total cost by dealing with total cost figures.

We now present other estimates that are refinements to the above simple approximation. The first of these is based on knowledge of which specific activities are to be carried out in the time period between ST_{50} and ST_{95}; where ST is the scheduled time as explained in Chapter 3.

2. *Activity-Based*—If it is known that certain project activities will be carried out after the *expected* project completion date, then the additional cost can be computed using specific costs for these certain activities. For example, if it is known in the home development project that the excessive time beyond 104 days would be expended on the driveway, sidewalk, shrubs, sod, finishing, and cleaning, then the rates for those activities would be used in computing the 95 percent cost.

At least an average of the last activities can be taken. If it is thought that, in the period between Day 104 and Day 110, that Activities 38–39–40–41 are the most probable ones that will be carried out and, therefore, the ones to incur cost, then the extended cost will be based on those activity rates. An average of those rates will be multiplied by the extended time and added to the expected cost in the six days from Day 104 to Day 110. Thus,

$Cost_{50\%}$ + average daily rate × 6 days = $Cost_{95\%}$

Therefore, in our home development project, the new cost will be

$39,130 + $213/day × 6 = $40,410

3. *Parallel Activities*

Another refinement would be to include extended parallel activities if there are near-critical activities.

The important point here is that the 95 percent probability cost can and must be calculated. Three methods are given. This is necessary because the expected cost has only a 50 percent certainty.

It is critical to closely estimate the cost that the company will incur in carrying out the development project. There are only so many dollars available in a company, and there are always many more project ideas presented than there are available funds.

The cost estimate is important as a primary element in the business plan.

The cost estimate is important as a control tool when the project is being carried out.

The top managers must decide which projects to pursue. This decision is based on company mission, competition, and available resources, including funds. Some companies spend 5 percent of sales on R & D, some 25 percent. Let's say a company has decided it can spend 10 percent of its sales of $50 million. That is, there are $5 million available in the coming year for R & D. Another $5 million should be planned for marketing research and marketing development. Additional funds will also be required for quality engineering, capital equipment for production, and other nonrecurrent expenses to be incurred.

The point is, only so much is available for new product development. It is crucial to closely estimate the cost that the company will incur in carrying out each of the proposed development projects. Only then can the top managers make an intelligent decision regarding which to pursue.

The cost estimate is important as a primary element in the business plan and as a control tool later on when the project is being carried out.

6 THE REPORT/PROPOSAL

INTRODUCTION

The procedures for planning and scheduling are now finished. We have completed Elements 1 through 6 of the *Overall Procedure*, which have been highlighted on page 174, and have performed as a team in carrying out concurrent project management. We now possess a thorough project plan, including final specifications, a project description, a plan network, a resource allocation, and a complete cost schedule.

It is time now to finish the preparation of a proposal and to submit it to higher management for approval and for project funding.

The report is a request for approval of the project just planned and scheduled and is a request for funding.

The seventh and final element of planning and scheduling is the report to higher management stating that the planning and scheduling are complete and giving the essential results. Every report should end with a request for a response. In this case, the request is for approval of the project just planned and scheduled and for funding.

174

OVERALL PROCEDURE

1. Clarify the *FUNCTIONAL SPECIFICATIONS*. Clarify the company objectives and the marketing objectives of the new product development. Study competition. Determine customer's needs.

2. Detail the product specifications. Identify and establish the cross-functional team. Study competition. Study the functional specification. Select the design concept. Detail all aspects of the physical product. Rely on experts. Write a *DETAILED PRODUCT DEFINITION*.

3. Describe the project. Elements 1 and 2, above, describe the product. This element, the *PROJECT DESCRIPTION*, describes the project of developing the product.

4. Generate the *PLAN NETWORK*. List all the tasks/ activities that compose your project. Prioritize them. Indicate after each activity the one or more other activities that immediately precede the one under study. Draw a rough network to illustrate the correct configuration.

5. Validate the *RESOURCE ALLOCATION*. Draw the work breakdown structure and assign the different talents to corresponding tasks. Obtain approval from the source managers for scheduled allocations.

6. Generate the *COST SCHEDULE*. Draw a cost curve as a function of time using the same time scale as used for the plan network and the resource histogram.

7. *REPORT* the plan to management and request funding.

THE REPORT/PROPOSAL

This report should consist of two main parts: (1) an executive summary of no more than 200 words, although 200 words is not a hard and fast rule, and (2) the back-up material that validates the summary.

The summary ought to be first and should be concise, filled with facts and bottom line statements. Everything that top management needs in order to approve the project should be included. Too many times, as a project manager or company executive, I have seen reports that have been too verbose with a paucity of hard evidence. That is, the written submission was too long and did not say enough with sufficient credibility to base a hard decision on. The company must select only a handful of projects out of all that are presented. The selection, in each case, is a difficult task, and some projects submitted will not be approved. The "pie" that must be cut and divided is only so big. Using our previous example, in Chapter 5, that "pie" is only 10 percent of sales, which is all that can be committed to product development in a given year.

Therefore, the report must exhibit credibility. Think. Given credibility, only one statement need be made: This project has been planned and scheduled, and the cross-functional team is ready to begin; therefore, please approve funding for carrying out the activities composing product development. Given credibility, management needs no more. At the opposite end of the spectrum of credibility, there is the report that is far too verbose and lacking in facts. We do not need to waste our time here considering such a poor report.

The back-up material that supports the summary can be the documents already prepared as the elements of the *Overall Procedure* were carried out. For example, when we completed the resource allocation, we ended with charts for them, Figures 4.8 and 4.9, that identified the various people and materials required. These charts can be attached to the executive summary as part of the report. The report or proposal should have the following features:

1. Should be addressed to a specific individual.
2. Should have a title page.
3. Should have a 200-word summary report stating that the planning and scheduling have been completed. It also refers to Elements 1 through 6 of the *Overall Procedure;* this is essentially a table of contents because these elements are included in the proposal. The whole proposal is paginated, so reference in the cover report to each element is by title and page number.

   - The 200-word summary should identify all the cross-functional team members and state the function of each: lead hardware engineer, senior mechanical engineer, software manager, and so forth.

   - The 200-word summary should also state who the project leader is by name.

   - It should state that the team has verified that all resources can be obtained on schedule. It should also state whether resources outside the project leader's department are required, and, if so, that approval from the various department and company heads has already been acquired.

   - The 200-word cover memorandum is an executive summary of the overall proposal.

- The 200-word executive summary should end with a request for funding.
- The report/proposal must be signed and dated.

It has been found that a 200-word summary is of adequate length.

The back-up material is the combination of elements resulting from following the *Overall Procedure*. These six documents should be attached to the Executive Summary in the same order in which they were generated. They are the following:

1. Functional Specification
2. Detailed Product Specification
3. Project Description
4. Plan Network
5. Resource Allocation
6. Cost

We now have completed the seventh and final step in the *Overall Procedure*, the report/proposal for funding.

We have run all these elements of the procedure, Elements 1 through 6, through the refinement defined in Chapter 4 and illustrated in Figure 4.4. We now have a set of refined elements, 1 through 6, to present as substantiation. Can we be more credible than that? No, not without going beyond the point of diminishing returns.

The proposal must be typed, neat, and written in good English. This writer has reviewed many, many project proposals, some with very good grammar and composition, some with very poor grammar and composition. True, the content is the most important feature of a proposal. However, even with excellent content, proposals have been rejected in industry because the English composition, grammar, and spelling were lacking in some way. The reviewer just could not understand the report well enough to decide whether to approve the project. Also, being careless in these areas leaves a bad impression because the reviewer has every right to believe that poor English indicates carelessness in carrying out the project.

A proposal carefully prepared in content and carefully expressed in English projects an image to the reviewer that indicates that great care will be taken in carrying out the project.

The report/proposal for the home development project is on the next page.

TO: (Name of supervisor)

HOME DEVELOPMENT REPORT
(Executive Summary)

The project of developing a home has been planned and scheduled. All of the objectives have been considered and clarified for all team members, who are the following:

Date

Project Leader–general contractor: (Name)_____

Framing subcontractor: (Name)_____

Plumbing subcontractor: (Name)_____

Electrical subcontractor: (Name)_____

Painting subcontractor: (Name)_____

A Functional Specification has been clarified and documented. See Page 3.

A Product Definition has been detailed and documented. See Page 4.

A detailed Project Description is presented on Page 5 of this report.

All the elements of good planning have been addressed. The project has been broken down into forty-five activities; these have all been prioritized and presented as either parallel or sequential activities.

The activities are shown in the Plan Network on Page 6. This is a graphic illustration: The sequencing and activity interrelationships are clearly shown and the critical path has been identified.

The resources required to carry out this project will be drawn from the appropriate subcontractors when necessary. Approval by the respective company heads has been obtained. The resource allocation is outlined on Page 7.

It is also expected that the project will be completed by Day 104. This expected completion time has a probability of 50 percent. Therefore, a second completion time has also been computed. It can be proven that there is a 95 percent probability of completing the project by Day 110.

It is expected that the total cost of the project will be $125,000. This expected cost has a probability of 50 percent. Therefore, a second cost has also been computed. The cost matching the 95 percent completion date is $135,000.

Your approval to carry out this project is hereby requested.

_____
(Name)
Project Leader
(Date)

7 MULTIPLE PROJECTS

INTRODUCTION

Many companies have several projects running simultaneously. Who among us has ever had the luxury of occupation with only one project? Some of us, yes, but not for long. If we have had this situation, it was probably because it was complex. But, almost by definition, our one project consisted of many subprojects. Many of these were more complicated than other whole projects we had previously completed. So, in a sense, we were really managing multiple projects even in this situation.

An individual may find it necessary to address issues on three or four projects within a given time period. His or her time must be shared between them before any one is completed. Also, it is found that other resources must be time-shared. Quite often engineering support services are divided between multiple projects, and the model shop, drafting personnel, and CAD workstations are also shared.

How, then, do we apply the foregoing material to this multiproject situation? How do we allocate and prioritize our time and other resources? How do we schedule these various projects and how do we control the schedule and costs of all projects?

We must take great care to prevent ourselves from working on too many projects simultaneously.

We should interject, at this point, some words of caution. We must take great care to prevent ourselves from working on too many projects simultaneously. Those who are responsible for running the business and assigning projects must be wary of spreading the limited resources too thinly. When this happens, projects do not finish on time. Even high-priority projects suffer schedule slippage when resources are too thinly spread. It is far better to apply resources to two or three projects, complete them, put the products on the market, properly support them, and, then, go on to the next projects.

We must insist on establishing a project priority with the top managers and demand that this priority not be changed without top-level consideration and direction. Once we begin a project and are well into it, there must be a major revelation in the market or a major technological breakthrough in another project for a change in priority to be justified. Many industry people say that management is always changing specifications and project priority.

An important part of project management is working with others within the organization to stabilize both the product definitions and project priority. Changing either devastates the time schedule and, therefore, the cost.

Deming's Point No. 7 is Institute Leadership. Leadership is the job of management.[1] Total Quality Management requires better management of specifications and priorities than we hear about from industry people.

It turns out that we can apply most of what we have presented in the foregoing chapters to the management of multiple projects. Let us draw some interesting parallels.

THE PROCEDURE

A good way of illustrating and managing multiple projects (consuming resources from the same pool) is to consider them all together and draw one large network diagram by connecting the projects by using dummy

---

[1] Mary Walton, *The Deming Management Method,* Putnam Publishing Group, New York, 1986.

activities. We rationalize this procedure by realizing that there is little difference between the large, complex project with many sub-projects and several small projects within a company.

> *A large, complex project is similar in many respects to many small projects drawing resources from the same pool.*

Basically, we will connect all the projects together with dummy activities at the appropriate points, then treat them as though they are different major activities within one large project. This use of dummy activities was alluded to in Chapter 2. Let us refer to this overall project within the company or other organization as a superproject. We now draw the parallels. Consider Situation A:

A. Every project is composed of many activities all interrelated. Each and every activity can be broken down into sub-activities.

Now, in the above paragraph, replace the word *project* with the word *company*. Replace the word *activity* with *project*. Replace *subactivity* with *activity*. The result is this:

B. Every company is composed of many projects all interrelated. Each and every project can be broken down into activities.

The point is this: The process and all the issues that concerned us in the foregoing chapters in dealing with the single project apply also to multiple projects.

We will follow the same *Overall Procedure* as before. We generate the functional specifications, detailed product specifications, project descriptions, plan networks, resource allocations, and cost schedules. The plan networks, allocations, and cost schedules for all the projects are combined at the next higher level of management into one management tool in each case. For example, when three projects draw resources from the same pool, we can combine all three project plan networks into one large plan network. Then resources can be allocated to this larger system and costs can, also, be scheduled for the combined, multiple-project situation.

The procedure is this:

1. Prioritize the projects
2. Connect the different projects together with dummy activities to indicate precedence
3. Allocate resources

THE PROCEDURE APPLIED TO THREE SIMULTANEOUS PROJECTS

We illustrate the recommended process for managing multiple projects by the following example. Consider a project consisting of these activities:

design system (DS)

preliminary design of product (PD)

construct prototype for feasibility testing (CP)

test feasibility (TF)

write operator's manual (OM)

design and develop the product (DDP)

design and develop the manufacturing process (DMP)

design and develop the manufacturing facilities (DMF)

design and develop the quality engineering procedures and equipment (DQE)

procure test equipment (TE)

procure components for company-wide prototypes (PC)

construct prototypes (CPs)

test prototypes (TP)

estimate cost (C)

carry out beta testing (BT)

go into production (Prod)

Figure 7.1 illustrates the project scheduled as a single project. The activity times are as shown adjacent to each activity arrow.

Now let us consider two additional projects that the organization is carrying out. They are basically the same as far as the activity definitions are concerned because most projects require system and product definition, prototype construction, and so forth. However, the activity times are somewhat different and are as listed in Table 7.1.

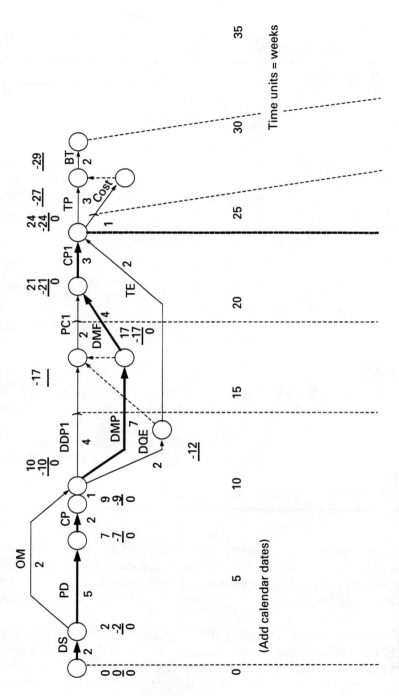

FIGURE 7.1 Plan Network for Project₁.

TABLE 7.1 Activity times for three projects.

ACTIVITY	PRODUCT $_1$	PRODUCT $_2$	PRODUCT $_3$
DS	2	2	2
DP	5	5	5
CP	2	2	2
TF	1	1	1
OM	2	2	2
DDP	**4**	**2**	**7**
PC	**2**	**2**	**2**
DMP	7	7	7
DMF	4	4	4
DQE	3	3	3
TE	2	2	2
CP	**3**	**4**	**3**
Cost	**1**	**1**	**1**
TP	3	3	3
BT	2	2	2

It is understood that if there were unlimited resources, all three projects could be run simultaneously without any dependency. But resources are limited, and there is dependency. If they were independent, then all three could finish in accordance with their own separate schedules, but they are not independent.

It is found, when scheduling the three projects, that computer-aided-design (CAD) talent must be shared. The design and development of the product relies heavily on CAD personnel and equipment. Also, the purchasing department must buy for all projects, not just one. Also, there is only one model shop, and it will construct all of the prototypes. The cost department estimates the cost of all new products. These resources and corresponding times have been set in **bold** type in Table 7.1.

On the basis of top-level business conditions, management has decided that the projects are to have the priority specified by the order in which they are considered here. That is, $Project_1$ has higher priority than the other two. $Project_3$ is the lowest priority project.

For purposes of illustration, let us assume that all three projects begin on the same day. Whether they do or not does not influence the procedure.

Figure 7.2 is the plan network for $Project_2$. We use subscripts to denote the activities that have different times from project to project. For example, in $Project_2$, DDP_2 is the design and development of $Product_2$. Because CAD resources are common, DDP_2 can not begin until after DDP_1 is complete. We draw a dummy activity dashed line from the end of DDP_1 in Figure 7.1 to the beginning of DDP_2 in Figure 7.2 to show that DDP_1 must be completed before DDP_2 can begin. Also, PC_2 can not begin until PC_1 is complete. CP_1 must be complete before CP_2 can begin. $Cost_1$ must be complete before $Cost_2$ can begin. We draw dashed lines connecting these activity endings in Figure 7.1 and the respective beginnings in Figure 7.2.

Resources for the other activities in all three projects are independently sourced. For example, the people who will write the operator's manual for $Project_1$ are different from those who will write the manual for $Project_2$, and, yet a different person or persons will write the manual for $Project_3$.

Figure 7.3 illustrates $Project_3$. There are similar dependencies between $Project_2$ and $Project_3$, the lowest priority project. Again, dashed lines between the two figures identify the dependencies.

It is interesting to note the effect limited resources have on the completion times of the two lower-priority projects. If there were unlimited resources, all projects would finish in thirty weeks, the length of the longest project, $Project_2$. However, it can be seen that limited resources have caused the finish of $Project_2$ to be at the end of thirty-three weeks and that of $Project_3$ to be at the end of thirty-six weeks.

We can find the critical path through the superproject by the same rules as before. In order to do so, we must first connect the starting events together with dummy activities, and connect the ending events in the same manner. We find the earliest event times and the latest times as in Chapter 3. We then learn, by applying the rules for identifying the critical path given in Chapter 3, that the critical path is DS_1–PD_1–CP_1–TP_1–DMP_1–DMF_1–CP_1–CP_2–CP_3–TP_3–BT_3.

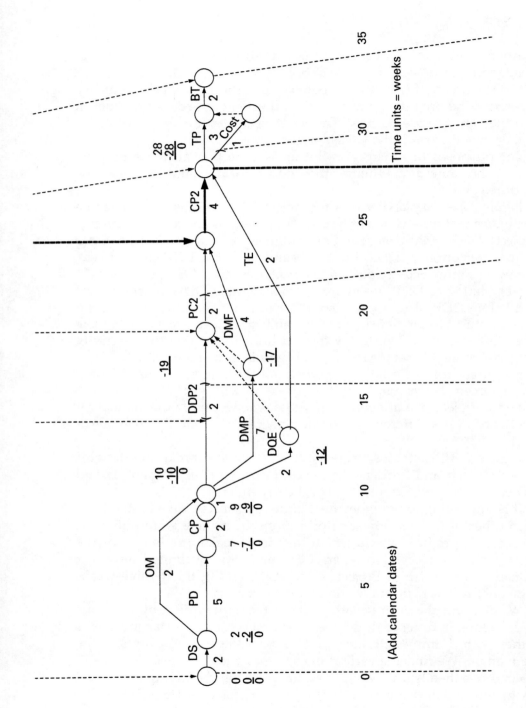

FIGURE 7.2 Plan Network for Project$_2$.

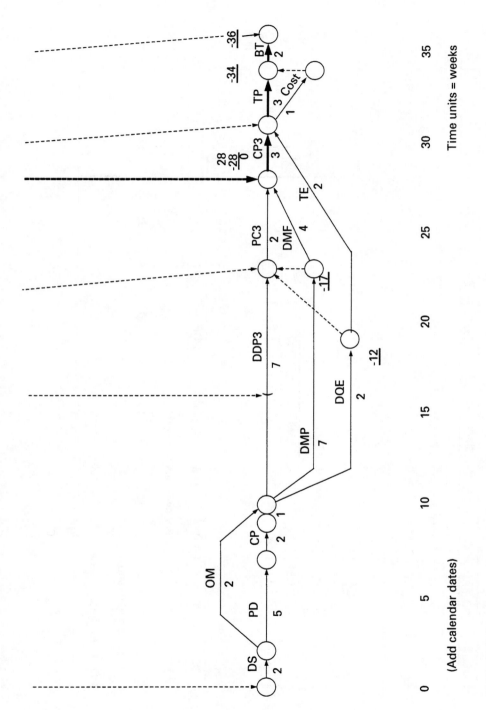

FIGURE 7.3 Plan Network for Project$_3$.

Time units = weeks

(Add calendar dates)

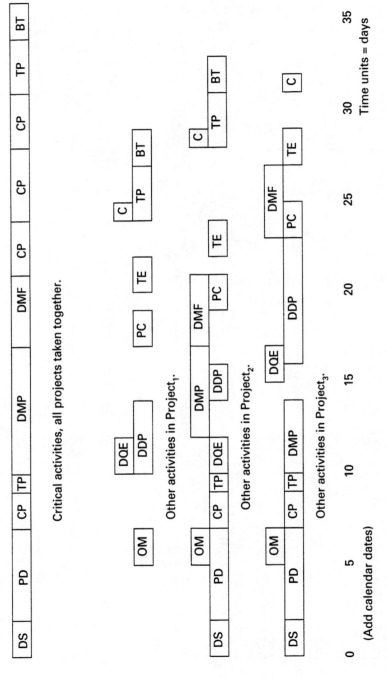

Critical activities, all projects taken together.

Other activities in Project$_1$.

Other activities in Project$_2$.

Other activities in Project$_3$.

Time units = days

(Add calendar dates)

FIGURE 7.4 Resource allocation for multiple projects.

The resource allocation chart is generated as before, and is illustrated in Figure 7.4.

The cost schedule is generated in the same manner as in Chapter 5. The project depicted in Figure 7.4 can be considered one large project, and the cost computed for each resource. The cost of materials is added to that for human resources to schedule the total cost.

8 CONCURRENT CONTROL

INTRODUCTION

It is stated in Chapter 2 that, by far, the best way to control a project is to carry out detailed, accurate, realistic planning at the outset. This *seems* so understandable and logical that it should not even be a subject of discussion at this point. However, there is a need to stress the importance of better planning and scheduling because many companies do not plan well enough, especially now when there is a strong trend away from cost-plus contracts toward fixed-cost, fixed-deliverables contracts. We should always plan excellently, in any case. However, fixed-cost, fixed-deliverables agreements *demand* excellent planning. I have taught a concurrent project management course many times, and have taken a survey by asking the question: "What do you wish to acquire from taking this course?" The results of the survey indicate that most participants need more training on planning and scheduling. In many cases, not enough is done before the beginning of the projects within the respective companies. Those managing them simply forge right ahead. There is strong indication that project control suffers because sound planning and realistic scheduling were not carried out, and most projects slip schedule. In all but a few cases, slippage was measured against schedules that were not realistic in the first place.

There are several excuses given by companies for not spending more time on planning. None are good enough, especially in these modern times of global competition. We make it a point to emphasize planning in this book. Every time I have taught a course or managed a project, planning is stressed and there is much dialogue with the participants, many of whom work in industry, concerning the elements of planning and scheduling. We plan and schedule right up to the point of diminishing returns. We convince ourselves that it is the best way to *control* the project later on. We know that without sound planning and realistic scheduling before the project begins, it is impossible to achieve necessary control of the project later on. Imagine if planning and scheduling were slip-shod—carried out to some extent but not detailed, accurate, or complete. This is the mode in which many companies operate, and the project management, as a discipline, suffers. Because the planning and scheduling were unrealistic at the outset, and, more to the point, too optimistic, the schedule begins to slip and continually slip. This follows by definition, because an overly optimistic plan, in fact, *implies* project schedule deterioration. It is almost impossible to correct an unrealistic schedule once it has slipped—for the same reasons it has slipped in the first place: The plan and the schedule were unrealistic. In fact, in this case, the words *unrealistic* and *impossible* are synonymous.

> *Without sound planning and realistic scheduling before the project begins, it is impossible to achieve good control of the project later on.*

The solution to this problem is to prevent it from slipping in the first place. It is absolutely imperative that the schedule be *realistic,* and, therefore, it is important to step back and assess the progress and status of the project management that we are carrying out.

As a team, check the design concept and all the figures, priorities, resources, and other aspects of planning and scheduling to make sure that we cannot do better (and that no one else can either), so that when we go before the review board, we succeed. We have given serious thought to each activity and have imagined carrying them out in the course of estimating the most likely, optimistic, and pessimistic time schedules. As a result of such good planning, it will seem to us and the review board that we have already successfully lived through the project. We must be well prepared; review boards can be brutal. But when our presentation for approval is highly organized, polished, concise, and credible, then they are very supportive. The reviewers want to fund projects and have responsibility for developing new products for the company. Therefore, when our project plan is sound, *we* will be funded and will work on *our* project—otherwise we will end up working on someone else's project and ideas—with our efforts at some lower level.

Let us assume that we have planned and scheduled better than anyone else can and that *we* are funded. Gaining approval was assured and we have attained concurrence at all levels of management. The team begins carrying the project out.

We must observe the progress and status of activity completion, and control the dynamics of carrying out the project. Observation and control are the subjects of this chapter. Before we study them, however, let us discuss one very important matter, which, if properly implemented, will lead to much better project control. It is the subject of decision making. Then, several additional suggestions will be made.

DECISION MAKING

It is necessary sometimes to make rational decisions when the choice between the alternatives is unclear. It always helps in these cases to

1. prioritize what is to be accomplished.
2. evaluate alternatives to generate overall value for each.
3. make a choice that balances benefits and risks.
4. sell the decision to others in the organization.

There are two parts to decision making:

A. The decision statement, in which we must establish objectives, classify objectives into the "must haves" and the "desirables," develop and compare alternatives, and consider adverse consequences.

B. The quantitative analysis, by which we make the decision. The following two pages gives a comparative analysis used successfully to determine which applicant to employ. Two absolute requirements for this method to result in a good decision are these:

1. *all* factors must be entered in the analysis
2. the *correct weight* must be assigned to each and every factor

It is assumed that accurate scoring is done. This can best be accomplished by three well-qualified people working together. In the case studied here (deciding which of several people to hire), ten applicants were interviewed, two were selected and three people interviewed, each of them separately. In addition, references were checked and all information was scrutinized and discussed by the three analysts prior to entering the score for each factor.

Tables 8.1 and 8.2 list all the factors critical to the position of drafting and documentation control manager. The weighting factors, scores, and total numerical evaluation of each issue are given. To the extent that *all* factors, the *correct* weight, and an *accurate* analysis of each issue is made, then the total numerical evaluation will lead to the correct decision.

TABLE 8.1 Decision Matrix I for recruiting.

SKILLS	SCORE	WEIGHT	TOTAL
Part numbers	10	8	80
Documentation control procedures and management	10	10	100
ECO management	8	10	80
Bill of Material management	10	10	100
Familiarity with computers	8	8	64
Software documentation	8	10	80
Drafting	6	5	30
Manufacturing documentation	5	8	40
Component familiarity			
PCB components	10	5	50
Power supplies, keyboards, etc.	6	5	30
Components peculiar to company	0	5	0
Coordination with Purchasing	10	10	100
Assembly training	2	5	10
Facilities maintenance	3	5	15
PERSONALITY			
Energetic	10	10	100
Cooperative	10	10	100
Influence others to cooperate	10	10	100
Detail oriented	10	10	100
"Perfectionist"	10	10	100
Hardworking	10	10	100
Giving to the company, vs. taking	8	10	80
			1459

TABLE 8.2 Decision Matrix II for recruiting.

SKILLS	SCORE	WEIGHT	TOTAL
Part numbers	10	8	80
Documentation control procedures and management	10	10	100
ECO management	6	10	60
Bill of Material management	10	10	100
Familiarity with computers	7	8	56
Software documentation	3	10	30
Drafting	8	5	40
Manufacturing documentation	8	8	64
Component familiarity			
PCB components	8	5	40
Power supplies, keyboards, etc.	9	5	45
Components peculiar to company	0	5	0
Coordination with Purchasing	10	10	100
Assembly training	10	5	50
Facilities maintenance	6	5	30
PERSONALITY			
Energetic	10	10	100
Cooperative	10	10	100
Influence others to cooperate	8	10	80
Detail oriented	10	10	100
"Perfectionist"	9	10	90
Hardworking	10	10	100
Giving to the company, vs. taking	6	10	60
			1425

Another example to illustrate where quantitative decision making would be beneficial is the case where the choice must be made between single and parallel processing in the design of a computer-based product. The factors to be included in the process would be those listed in the left hand column of Table 8.3.

ADDITIONAL SUGGESTIONS

1. Commitment. There must be a real sense of commitment on the part of each project participant in order for good project control to exist.

TABLE 8.3 Decision Matrix for deciding which processing mode to use.

	SINGLE PROCESSOR	PARALLEL PROCESSORS
Overall Computing Speed		
Hardware Development Cost		
Software Development Cost		
Hardware Reliability		
Software Reliability		
Hardware Cost		
Others?		

The plan network has been formulated and drawn up by the whole cross-functional team. The camaraderie and rapport that has been established has led to a definite responsibility on the part of each team member and other people who will be carrying out activities. This commitment can be underscored by having the name of the individual in charge of carrying out the activity in each case entered adjacent to the activity on the plan network. See Figures 4.11 and 4.12. This highlights the responsibility and enhances the sense of commitment. If possible, give each person obligated to carrying out an activity or activities a copy of the plan network. If the project is too confidential to do this, then give each person a written schedule of when he or she must fulfill the agreements.

2. Prevent people from being "spread too thinly." If we allow ourselves to be involved in too many projects, it will be impossible to control any of them. Establish the priority of projects with top management. They will agree that it is far better to concentrate our efforts and other resources on the top one or two than to try to keep five, ten, or twenty projects running neck and neck. We must not tolerate priority changes once the subject project begins. If this seems too naive, consider the alternative. Does any member of the cross-functional team want the schedule to be devastated? Does top management? The answer is no. Once we are well into the project, we must not permit anyone to shift priorities so as to significantly affect the schedule.

3. The Product A, Product B concept. In some projects, it may be better to develop a relatively simple product that meets the specifications and to get it on the market; then develop an upgrade to this product and put it out at some later time. This is a smarter alternative than to wait significantly longer until the final version is developed, with more "bells and whistles." In another instance, Product A may be developed faster by using more outside-purchased components than the final version will have. In this final version (Product B), more of the components may be developed and manufactured in-house to keep the cost down and to have more control over the manufacturing schedules. In some cases, both products may be kept in the product line. In other cases, Product B will replace Product A. Care must be taken, in any case, to provide high quality in both products.

OBSERVATION

In order to properly control the project, we must first observe it. Let us consider a parallel situation. To properly drive an automobile along a highway, we must first observe its position and direction in comparison to the centerline of our lane. We must also observe our speed. All the factors that are important to the success of the project, in this case the automobile trip, must be acknowledged.

Observation is evaluating the current status of the project. Each activity in the overall project must be evaluated. This involves several key measurements of the status:

A. The quality of performance—how well has the task or activity been carried out?

Example: In an electromechanical/hydraulic pressure transducer development project, consider the activity of designing the mechanical part of the transducer:

1. Have all aspects of the functional specification for the mechanical part of the design been addressed?
2. Have all aspects of the detailed product specification for the mechanical part of the design been addressed?

Are there any other requirements, such as FDA approval, electrical codes, or intrinsic safety aspects, that must be met—that may have been overlooked when the specifications were approved before?

Has experienced, ingenious talent been applied to this task with excellent results—even in the conceptual design? During the development and product engineering phases?

B. Is the activity complete? As a measure of completeness, consider the following issues:

1. Has sound engineering been carried out?

2. Has the product been well simulated, prototyped, and tested? Has a statistically significant number of prototypes been tested and qualified?

3. Is the design well documented?

4. Is the task complete enough that we will never have to return to it later in the project?

   (As a test, imagine that if we do have to return to the task and re-do some part of it, because we were not careful enough when we carried out the activity, we will have to pay for it with our own personal money and time —not the company's!)

C. Did we complete the activity on time?

D. Did we complete the activity within budget?

E. Did we meet any other possible requirements?

This process can be viewed as measurements of signals going into an AND gate in a logic network. A through E above are the AND gate inputs as illustrated in Figure 8.1. If, and only if, all inputs are true, the output is

FIGURE 8.1 Observation logic.

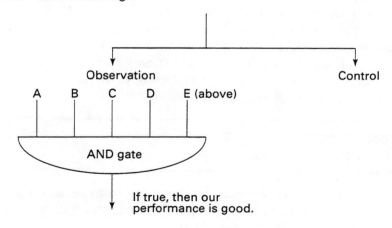

true; that is, we are in good shape with the status of the project if and only if we can answer yes to all the questions, A through E.

CONTROL

The process of observation and control can be compared with the dynamic process in a control system. Most people are familiar with the concept of feedback control because we are all subject to many control systems in our everyday lives: the home heating system, blood pressure and body temperature, the dilation and contraction of the pupil in the eye in the presence of changing light intensity, driving an automobile along the centerline of a highway lane, cruise control, and our own weight control. So we understand the broad dynamics of feedback control systems, and it is useful as an analogy, here. To assist the reader in comprehending control system dynamics, we define some examples. In the home heating system, one hypothetically sets the thermostat to 65° F. The action taken at this point is to establish the desired temperature setting. Now, if the actual temperature is only 60° F, then the heating plant turns on and the living space is heated until it is 65° F. During this process, the actual temperature is continuously "observed" by the thermometer component of the thermostat and compared with the setting. When the actual temperature reaches the desired temperature of 65° F, then the heating plant is turned off until it is needed again. There is both observation and control exercised in this dynamic process.

Consider next the control system dynamics in the process of driving an automobile. The desired input is that of the vehicle moving down the centerline of the chosen lane. Its actual movement, however, is not exactly in alignment with the centerline, but the centerline is used as an average, with minor deviations from it. The driver observes the actual performance and makes small corrections with the steering wheel so as to minimize the deviations. Again, there is both observation and control in this dynamic control process. Project management, as in all closed-loop control systems, requires feedback control, by definition.

> *Project management, as in all closed-loop control systems, requires feedback control.*

All projects have the potential for deviating from perfect control, just as any other control system does. Much of this chapter will be devoted to solutions when there is project deviation.

The input to the control system, in the case of project management, is the set of requirements detailed in the specifications, the plan network, the resource allocation, and the cost schedule. These lead to quality, schedule control, and cost control.

The output from the control system is the set of actual *results* in terms of the quality of performance, the schedule, and the cost. That is, the output is the actual status with regard to the degree of project completion compared with what the schedule stated for the date of measurement. It is the status with respect to how much of the budget has been spent compared with the cost schedule, and with respect to any other factors contained within A through E as defined previously as they were observed at any given point in time. See Figure 8.2 which illustrates our project control system.

Another way of viewing the observation function is illustrated in Figure 8.3.

If A, B, C, D, and E are not all true, then why? What is the amount of control system error? Who is responsible and best qualified to compensate for the error with the appropriate control system plan? Based on the specific control system dynamics of our new product development project, what is the best type and amount of compensation?

The cross-functional team controls the dynamics of carrying out the project by way of each individual or group of individuals performing their respective activities. Again, the team will have to be especially creative in effecting the right type and correct amount of compensation. See Chapter 1.

The controlling function is the process of regulating and directing the progress of the project so that it will be completed on or before the scheduled time of the plan network finish event. (For example, we may target the 95 percent time.)

FIGURE 8.2 Project control system.

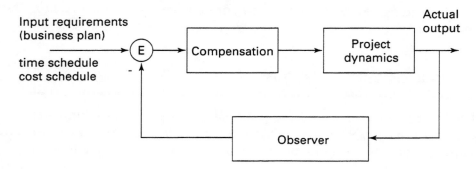

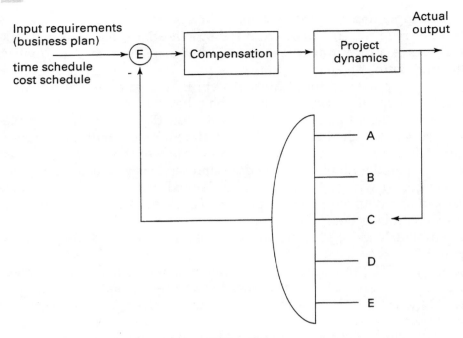

FIGURE 8.3 Project observation/control system.

The best way of controlling cost is to keep the project on the time schedule—then the cost schedule will most likely be satisfactory.

The key to controlling the project is preventing trouble before it starts by making use of the capability of the plan network to predict or forecast trouble areas. The plan network was very important for us as a planning and scheduling tool; now it is vital to us as a control tool.

Overall, the best way of controlling a project is to plan well and schedule realistically at the outset. On the other hand, we recognize that there will sometimes be deviations from the plan. In these cases, we can compensate by adding resources, if available, or by releasing Product A, first, then Product B, as discussed previously, or by working longer hours.

REPORTING PROGRESS AND STATUS

Having properly observed the progress and status, and taking adequate control of the project, it is now time to report the findings to all those individuals who have responsibility for it. Who are these people that we refer to? First, the cross-functional team that did the planning and scheduling, and who are now controlling the project, will want to document and record the progress and status against the plan network, the time schedule, and the cost schedule. Secondly, the immediate supervisors of each of the respective cross-functional team members will want to keep abreast of the welfare of the project that their people are working on. Thirdly, the people in the relevant chains of command within the company will need to know the progress and status of the new product development in order to continue the business planning and controlling at the highest levels within the company. People responsible for materials management, marketing, production, capital equipment procurement, and human resource allocation will all need to monitor the status of the new product development in order to be ready at the appropriate times to market, produce, and distribute the product. The window of new product opportunity is usually small. Advancing technology and strong competition drive rapid new-product evolution. Therefore, the new product must be ready when the market calls for it.

Why not use the same documents generated during the planning and scheduling phases for reporting the progress and status of the project?

Consequently, it is necessary to put in place and maintain a suitable reporting system. But there is no need to generate a whole new set of documents for this. Why not use the same ones produced during the planning and scheduling phases for reporting the progress and status of the project?

The plan network (which by now has probably been programmed into a computer for rapid updating and plotting) can very easily and effectively be used as a reporting medium, as can the cost curve generated as a part of the planning process. Management, including the cross-functional team project leader and higher management, needs to know whether the project is on schedule, and, if not, whether the project is ahead or behind, and by how much, and what the cross-functional team has done about it if there has been significant slippage. Also, the actual cost needs to be known in comparison with the original cost schedule, because the current company forecast and budget are based on this original schedule.

REPORTING ACTIVITY PROGRESS

The plan network is an accurate documentation and reporting tool because it is very comprehensive and it is easily understood. After all, the cross-functional team is the group most interested in the project's progress and status, because the team is working on the project every day and has everyday responsibility. And, after all, the cross-functional team that is carrying out the project is the same one that generated the plan network during the planning and scheduling phases. So the team is already intimately familiar with the plan network; this makes the plan network a reporting tool that is easily understood by them. Also, the next level of supervision in the company is familiar with the network, because they gave their concurrence with the elements of the network toward the end of the planning and scheduling phases. Consequently, the supervisors are familiar, a priori, with the plan network. This is another reason for using it as a reporting tool. Third, top management is already familiar with the network, because they were involved in the final approval for funding of the project. The plan network was prepared and presented as a part of the funding request package. Basically, the network gives a good pictorial illustration of the overall project. It shows how all the activities are interrelated and it highlights the critical path, as well as the near-critical paths. It is on a linear time scale (the calendar), and it is on the same time scale as the resource allocation and cost schedule. So, there are good reasons for using the plan network for documenting and reporting the progress and current status of the project at appropriate intervals during the execution of the project after it has been approved. The plan network should be limited to forty or fifty critical path activities, and no more than 100 activities all together; otherwise, it becomes too cumbersome for a reporting tool.

Managers at all relevant levels need to know of the current status of the project, including actual activity progress and actual cost to date.

Figure 8.4 is an illustration of the first part of the new product development project, and it includes the beginning twenty activities. At the initial design review meeting, held on Day 10, it was found that the first five activities had been completed. This is indicated by crosshatching those five activities on the plan network. At the next meeting, held on Day 17 when

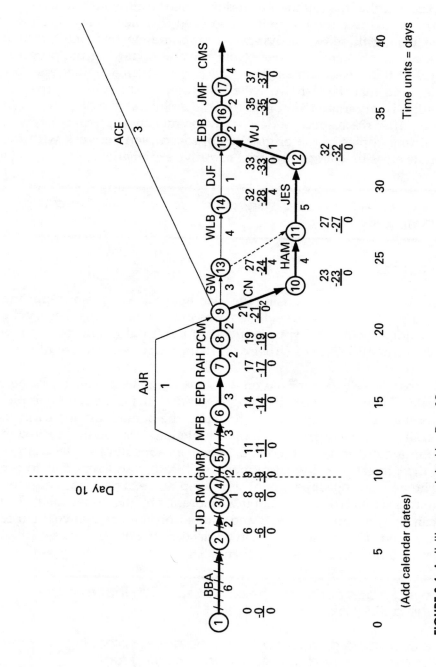

FIGURE 8.4 Activities completed by Day 10.

it was expected that the foundation and basement wall would be finished, it was found that two more activities had been completed. This is shown by crosshatching these two activities. At the same meeting, the first five activities are double crosshatched. Now, anyone reviewing the progress and status of the project can tell at a glance, that seven activities are now completed and that the last two were also completed since the previous meeting (if it is important to know that). This is illustrated on Figure 8.5. In both cases, after the respective design review meetings, a report is given by submitting copies of the originally approved plan network with the appropriate cross-hatching and a brief narrative explanation.

The plan network is a very good documentation and reporting tool.

REPORTING COST

The reasons given above for using the plan network as a major reporting tool also apply to the cost curve. It, too, was prepared and finalized by the cross-functional team, and was included in the funding approval package. Therefore, all levels of supervision and responsibility are already familiar with it.

The cost curve was finalized on the same linear time scale as the network scale. Therefore, when the progress and status of activity completion are reviewed periodically by the team, the actual cost incurred to date is easily qualified against activity completion. It is usually tracked in the accounting office. The true cost is known as a result of actual charges against the project, and the accountants will know and report it to the team. The accounting department will keep score. The team can then compare this actual cost with what it was forecast by the project plan. The actual can then be superimposed and plotted on the original cost curve for comparison. The actual curve can be dashed; the original curve solid, or different colors can be used. See Figure 8.6.

The team can compare the actual cost with that forecast in the project plan.

The cost curve generated during the project planning phase can be used as a significant reporting tool.

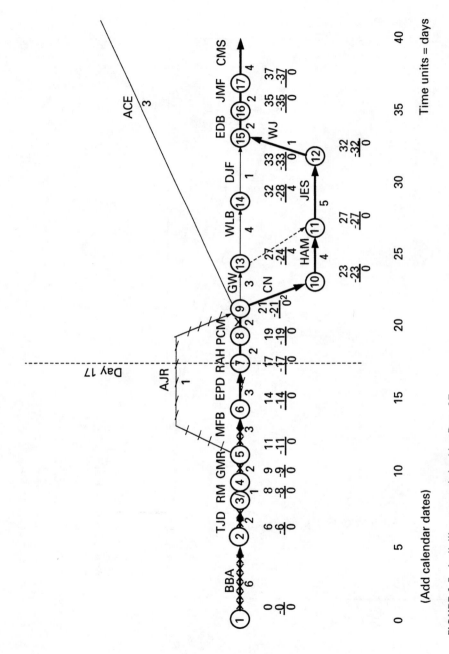

FIGURE 8.5 Activities completed by Day 17.

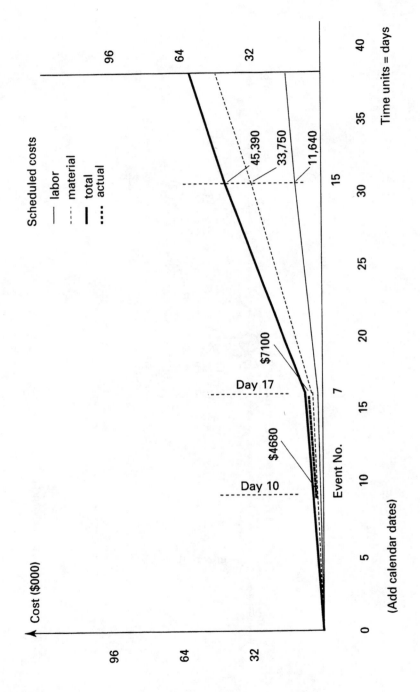

FIGURE 8.6 Cost for development of home.

The reader is reminded that both the cost schedule that is expected (50 percent probability) and the one matching a 95 percent probability of completion by the corresponding date are presented on the official cost schedule. Consequently, the project team, and the other company managers, have before them a complete picture regarding actual costs versus those forecast. That is, the actual cost can not only be compared with the 50 percent prediction but also with the more probable costs that were also in the original approval package. Again, just as there is a 50 percent chance of overrunning the expected time schedule, there is a 50 percent chance of overrunning the expected cost schedule. However, there is only a 5 percent chance of going beyond the 95 percent time and cost schedules originally submitted in the approval package. Figure 8.6 gives the actual cost as compared with the scheduled cost.

What are appropriate reporting intervals for the project? How frequently should the project leader report on the status of the project? Should there be a hard and fast reporting schedule? To whom should the project leader report? What should the elements of the report be? Should a response be requested? These and other questions will be answered in the next section.

FREQUENCY OF REPORTING

Appropriate intervals for reporting the progress and status of the project depend on its nature, the characteristics of the market, and on the management styles of the project leader, the next level of supervision, and top management. In a fast moving, highly competitive marketplace, and with a project that is intensive and incurring expenditures at a high rate, reporting must be relatively frequent.

For example, we might decide to review the home development project every five days, every month, or at certain predetermined milestones. These milestones might be at the expected completion dates of predefined activities. We might schedule review meetings every ten days and *also* at the expected completion dates of certain activities such as:

1. completion of foundation and basement walls

2. completion of framing and plumbing

3. completion of roof and basement floor

4. completion of wiring, exterior windows and doors, and lawn

5. completion of interior walls and doors

6. completion of final activities

In this example, we choose these particular milestones because they are major completion events, and our cost analysis shows that each successive milestone is scheduled at about the time that another $10,000 of labor is to be incurred. That is, the team will hold another formal review about the time another $10,000 is to be spent. In this example, it is about the time another 25 percent of the total labor cost is incurred. Management might use a guideline such as 20 percent or 25 percent. In addition, the review at ten day intervals is about 10 percent of the 104 day total. The reviews may determine that the team should study actual costs at the activity milestones such as defined above. Or the team may decide that costs, and activity progress, will be studied at the ten day points, as well as at the major milestones. Most companies do not keep a close enough vigil on activity progress and incurred cost. I recommend that the cross-functional team and management define at the outset of the project what the review points will be. This should be based on criteria as discussed above. Then, the team will follow this plan throughout the project.

> *Most companies do not keep close enough vigil on activity progress and incurred cost.*

Depending on the nature of the project, the market, and preferred management styles within the specific company, reporting to higher levels will take place at the milestones only, or at both the milestones and at periodic time intervals, such as every ten days or once a month. The project will be sufficiently controlled when all the people in the company who have interests in it communicate well and are productive together in overall project management.

In accordance with the plan, the project leader will, at the conclusion of each specified review meeting, report progress and status to the next level of supervision. The elements of the report are activity completion, actual cost (including labor and material), comparison of actual time and cost with scheduled time and cost, and any significant differences from the plan not only in schedules but also in product performance.

If approval for extra resources, such as additional people, funding for overtime, test equipment, or outside services is needed, then a response should be requested. Whether one is necessary depends if the cross-functional team feels it requires top management intervention. The dynamics of project control should be such that the people within the company who have the authority and responsibility to participate in the control of the project should always be communicating with upper management. Thus, any significant deviation from the plan will be known to all long before any matter becomes critical.

Finally, the team must analyze the project status so that the leader can state clearly in the report that the actual status of time and cost is compared with both the 50 percent expectations and the 95 percent schedules.

9 CASE STUDIES

INTRODUCTION

As stated in Chapter 1, four case studies will now be presented. All are new product development cases that are directly related to concurrent engineering. Three are success stories, the other is the result of improper management and is presented only for comparison. One of the success stories is taken from a conference attended in 1989, while the remaining three are direct experiences from companies where I was employed.

COLUMN GAGE

Several years ago Federal Products Corporation, a unit of the Esterline Corporation, undertook the development of a microprocessor-based instrument to display the measurements of thickness and diameters in quality control gauging. Several design concepts had been explored, and an industrial designer had been consulted on shape and color. The more traditional development approach of first carrying out design in R & D and then process development, later, by the operations people was planned. In fact, a prototype and cost study were made. It soon became

210

clear that it would be highly beneficial for a more integrated, cross-functional team to work concurrently on the overall development.

It was felt that a major change in the packaging concept was required, along with a complete reexamination of all the components and their respective needs. There were many subassemblies and fasteners, and it was determined that the number of each could be significantly reduced. The housing was a previously molded plastic design. It was thought that if aluminum extrusions were used instead, slip fits could be designed in, eliminating the need for fasteners. At the same time, this would provide a more rugged enclosure that would be more protective of the circuitry against static, and the product would be more radiation free.

A team consisting of the marketing product manager, the R & D designer, a manufacturing engineer, and an assembler was formed, and the concurrent engineering project was underway.

The fundamental objective was to improve reliability and cost. An important result was a modularized construction, partitioning the product at the minimum interconnection boundaries. This facilitates manufacturing and testing, in addition to inventory control of subassemblies, which is sometimes found necessary because of long lead-time deliveries of certain components. These subassemblies can be tested and stored as high-quality assemblies ready for integration into the final product.

The component count in the new column gauge was reduced by 25 percent, while the cost was reduced by 20 percent, and the assembly time by 15 percent. The development project was then completed and a mean-time-before-failure (MTBF) analysis on the resulting product indicated an MTBF of 18,600 hours. The new column gauge has been on the market for several years with good results.

DIAL BORE GAUGE

Another product, a dial bore gauge, consisting of a precision mechanism for gauging cam bores, cylinder bores, and other diameters was developed at Federal Products Corporation. The readout was a mechanical dial indicator. This product was not electrical, nor electronic at all, and should have been relatively easy to develop because fewer types of engineering disciplines were required.

However, the sequential method of carrying out the R & D, followed by the manufacturing engineering and quality control, was attempted. The design concept was quite unique; in fact, a patent with thirty-two claims was allowed. The problem was that manufacturing was not brought into the development process early enough for the operations people to become familiar with the project and to begin the manufacturing process development on a schedule leading to substantial profits.

From a company earnings standpoint, this lack of team integration resulted in an extremely long lead time to market. Another significant drawback was that a product champion was not designated and empowered by management.

LASER SURGICAL SYSTEM

I was recently retained as a consultant by Surgilase, Inc. to upgrade the management of new product development. Surgilase provides instrumentation systems to hospitals and surgeons for performing corrective procedures by laser surgery. These techniques are increasing in number and are now in widespread use by orthopedic surgeons and neurosurgeons, as well as by other specialists.

It is important to control the laser power and also the duration of the laser pulse. The system is microprocessor-based and depends heavily on competent software development as well as sound hardware engineering. Quality assurance is, of course, mandatory, and overall approval by the Food and Drug Administration must be obtained on all system designs.

A product had previously been developed, and the development project had cost $300,000 over a two-year period. The sequential development approach had resulted in significant redesign and excessive prototyping costs. Subsequently, a whole new concept of team-building and consensus management was implemented—structured around concurrence in functional and detailed product definition and concurrent engineering.

Members of the team established at that time were (1) the Marketing product manager, (2) the manager of Software and Electrical Engineering, who was designated as project leader, (3) the manager of Operations, which included Manufacturing Engineering and Production Control, (4) the manager of Quality Assurance and Customer Service, and (5) the manager of Mechanical Engineering. This team was commissioned and a letter asking for the support, whenever necessary, by the remainder of the company was written by the company president. This was recommended as a measure to empower the team. They met regularly, and developed a new laser system to meet the same functional specifications the previous system had with one exception; the new one had to be designed for outpatient procedures, as well. This was an important feature and pointed out an additional need for cross-functional teamwork. Even concurrence in the specifications is a necessary part of concurrent engineering. The need for outpatient service should have been addressed in the first design. It was a principal reason for having to carry out the new development project. The new system was developed in ten months at a cost of $125,000.

IBM PROPRINTER

The last case is a comparison between the IBM Selectric computer printer designed years ago and the recently designed IBM Proprinter. This comparison was presented by Ray H. Reichenbach at an SME conference, "Simultaneous Engineering: Making it Work," Dearborn, MI. It was reported that the Selectric had 2,700 parts, 140 adjustments and required an average of one service call per year. The Proprinter, a result of concurrent engineering, has sixty-two parts, no adjustments, and the trend indicates one service call every five years.

BIBLIOGRAPHY AND REFERENCES

1. *Management of Technology and Innovation: Strategic Issues Program,* Fuqua School of Business, Duke University, Durham, NC, October 5–10, 1986.
2. *R & D/ Technoloqy: Key Issues for Management,* New York: Conference Board, February 29–March 1, 1988.
3. *Fourth International Conference on Product Design for Manufacturability and Assembly,* Institute for Competitive Design and Boothroyd-Dewhurst, Newport, RI June 5–6, 1989.
4. *Simultaneous Engineering: Making It Work,* Society of Manufacturing Engineers, Dearborn, MI, June 8, 1989.
5. *Managing Multi-Functional Development Teams,* Manufacturing Institute, New Orleans, LA, September 25–26, 1989.
6. Aguayo, R.: *Dr. Deming,* Simon & Schuster, New York, 1991.
7. Badiru, A. B. and G. E. Whitehouse: *Computer Tools, Models and Techniques for Project Management,* TAB Books, New York, 1989.
8. Balderston, J., P. Birnbaum, R. Goodman, and M. Stahl: *Modern Management Techniques in Engineering and R & D,* Van Nostrand Reinhold, New York, 1984.
9. Boyett, J. and H. Conn: *Workplace 2000,* Dutton, New York, 1991.

10. Carter, D. E. and B. S. Baker: *Concurrent Engineering,* Addison-Wesley, Reading, MA, 1992.

11. Cleland, D. I.: *Project Management: Strategic Design and Implementation,* TAB Books, New York, 1989.

12. Cleland, D. I. and W. R. King: *Project Management Handbook,* Van Nostrand Reinhold, New York, 1988.

13. Cohen, H.: *You Can Negotiate Anything,* Bantam Books, New York, 1982.

14. Conger, J. A.: *Learning to Lead,* Jossey-Boss, San Francisco, 1992.

15. Covey, S. R.: *The Seven Habits of Highly Effective People,* Simon & Schuster, New York, 1990.

16. Craig, A. T. and R. V. Hogg: *Introduction to Mathematical Statistics,* MacMillan, New York, 1978.

17. Crawford, C. M.: *New Products Management,* Richard D. Irwin, Homewood, IL, 1983.

18. Gardner, J. W.: *On Leadership,* Free Press, New York, 1990.

19. Gido, J.: *An Introduction to Project Planning,* Industrial Press, New York, 1985.

20. Gilbreath, R. D.: *Winning at Project Management: What Works, What Fails and Why,* John Wiley, New York, 1986.

21. Grant, E. L.: *Statistical Quality Control,* McGraw-Hill, 1964.

22. Gray, C. F.: *Essentials of Project Management,* Petrocelli Books, Princeton, NJ, 1981.

23. Hajek, V. G.: *Management of Engineering Projects* McGraw-Hill, New York, 1984.

24. Halbertsam, D.: *The Next Century,* William Morrow, 1991.

25. Harrison, F. L.: *Advanced Project Management,* John Wiley, New York, 1981.

26. Hartley, J. R.: *Concurrent Engineering,* Productivity Press, Cambridge, MA, 1992.

27. Harvard Business Review Staff: *Managing Projects and Programs,* Harvard School of Business, Cambridge, MA, 1989.

28. Harvard Business Review Staff, *Project Management,* Harvard School of Business, Cambridge, MA, 1991.

29. Hiam, A.: *Closing the Quality Gap,* Prentice-Hall, Englewood Cliffs, NJ, 1992.

30. Hicks, T. G.: *Successful Engineering Management,* McGraw-Hill, New York, 1966.

31. House, R.: *Human Side of Project Management,* Addison-Wesley, Reading, MA, 1988.

32. Kerridge, A. E. and G. H. Vervalin: *Engineering and Construction Project Management,* Gulf Publishing Company, Houston, 1986.

33. Kerzner, H.: *Project Management,* 4th ed., Van Nostrand Reinhold, New York, 1992.

34. Kerzner: *Project Matrix Management Policy and Strategy*, Van Nostrand Reinhold, New York, 1984.

35. Kouzes, J. M. and B. Posner: *The Leadership Challenge*, Jossey-Bass, San Francisco, 1987.

36. Lineback, L. K.: *Being the Boss*, IEEE Press, New York, 1987.

37. Magaziner, I. and M. Patinkin: *The Silent War*, Random House, New York, 1989.

38. Moder, J.: *Project Management with CPM, PERT and PRECEDENCE Diagramming*, Van Nostrand Reinhold, New York, 1983.

39. Morris, W. G. and G. H. Hough: *The Anatomy of Major Projects: A Study of the Reality of Project Management*, John Wiley, New York, 1988.

40. Nyers, M. S.: *Every Employee a Manager*, McGraw-Hill, New York, 1970.

41. Peters, T. J.: *In Search of Excellence*, Warner Books, New York, 1982.

42. Phillips, D. T.: *Lincoln on Leadership*, Warner Books, New York, 1992.

43. Phillips, D. T., A. Ravindran, and J. J. Solberg, *Operations Research*, John Wiley, New York, 1976.

44. Richmond, S. B.: *Statistical Analysis*, Ronald Press, New York, 1964.

45. Rosenau, M. D.: *Project Management for Engineers*, Van Nostrand Reinhold, New York, 1984.

46. Sholtes, P. R.: *The Team Handbook*, Joiner Associates, Madison, WI, 1988.

47. Silverman, M.: *The Art of Managing Technical Projects*, Prentice-Hall, Englewood Cliffs, NJ, 1987.

48. Spinner, M. P.: *Elements of Project Management: Plan, Schedule and Control*, Prentice-Hall, Englewood Cliffs, NJ, 1981.

49. Stephanou, S. E.: *Management*, Daniel Spencer, Malibu, CA, 1981.

50. Stephanou, S. E. and M. M. Obradovitch: *Project Management*, Daniel Spencer, Malibu, CA, 1985.

51. Stuckenbruck, L. C.: *Implementation of Project Management: The Professionals Handbook*, Addison-Wesley, Reading, MA, 1981.

52. Thomsett, R.: *People and Project Management*, Prentice-Hall, Englewood Cliffs, NJ, 1986.

53. Walton, M.: *The Deming Management Method*, Putnam, New York, 1986.

54. Westney, R. E.: *Managing the Engineering and Construction of Small Projects*, Marcel Dekker, New York, 1985.

55. Williams, A.: *Pushing Up People*, Park Lake, Doraville, GA, 1984.

INDEX